NASA STI Program ... in Profile

Since its founding, NASA has been dedicated to the advancement of aeronautics and space science. The NASA scientific and technical information (STI) program plays a key part in helping NASA maintain this important role.

The NASA STI program operates under the auspices of the Agency Chief Information Officer. It collects, organizes, provides for archiving, and disseminates NASA's STI. The NASA STI program provides access to the NASA Aeronautics and Space Database and its public interface, the NASA Technical Report Server, thus providing one of the largest collections of aeronautical and space science STI in the world. Results are published in both non-NASA channels and by NASA in the NASA STI Report Series, which includes the following report types:

- TECHNICAL PUBLICATION. Reports of completed research or a major significant phase of research that present the results of NASA Programs and include extensive data or theoretical analysis. Includes compilations of significant scientific and technical data and information deemed to be of continuing reference value. NASA counterpart of peer-reviewed formal professional papers, but having less stringent limitations on manuscript length and extent of graphic presentations.

- TECHNICAL MEMORANDUM. Scientific and technical findings that are preliminary or of specialized interest, e.g., quick release reports, working papers, and bibliographies that contain minimal annotation. Does not contain extensive analysis.

- CONTRACTOR REPORT. Scientific and technical findings by NASA-sponsored contractors and grantees.

- CONFERENCE PUBLICATION. Collected papers from scientific and technical conferences, symposia, seminars, or other meetings sponsored or co-sponsored by NASA.

- SPECIAL PUBLICATION. Scientific, technical, or historical information from NASA programs, projects, and missions, often concerned with subjects having substantial public interest.

- TECHNICAL TRANSLATION. English-language translations of foreign scientific and technical material pertinent to NASA's mission.

Specialized services also include organizing and publishing research results, distributing specialized research announcements and feeds, providing information desk and personal search support, and enabling data exchange services.

For more information about the NASA STI program, see the following:

- Access the NASA STI program home page at *http://www.sti.nasa.gov*

- E-mail your question to help@sti.nasa.gov

- Fax your question to the NASA STI Information Desk at 443-757-5803

- Phone the NASA STI Information Desk at 443-757-5802

- Write to:
  STI Information Desk
  NASA Center for AeroSpace Information
  7115 Standard Drive
  Hanover, MD 21076-1320

NASA/TM–2013-218032
NESC-RP-09-00605

Development of Autonomous Aerobraking

Phase 2

Daniel G. Murri/NESC
Langley Research Center, Hampton, Virginia

National Aeronautics and
Space Administration

Langley Research Center
Hampton, Virginia 23681-2199

August 2013

Acknowledgments

The following were significant contributors to this effort:

Jill L. Prince; Team Lead, LaRC

Richard W. Powell; NESC Deputy Lead, Flight Mechanics TDT, AMA

Alicia M. Dwyer Cianciola; LaRC Simulation, LaRC

The use of trademarks or names of manufacturers in the report is for accurate reporting and does not constitute an official endorsement, either expressed or implied, of such products or manufacturers by the National Aeronautics and Space Administration.

Available from:

NASA Center for AeroSpace Information
7115 Standard Drive
Hanover, MD 21076-1320
443-757-5802

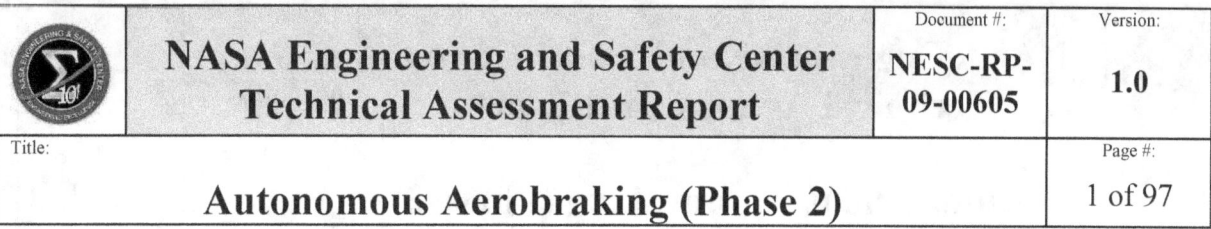

Development of Autonomous Aerobraking
(Phase 2)

June 13, 2013

NESC Request No.: TI-09-00605

Report Approval and Revision History

NOTE: This document was approved at the June 13, 2013, NRB. This document was submitted to the NESC Director on July 30, 2013, for configuration control.

Approved:	*Original Signature on File*	7/30/13
	NESC Director	Date

Version	Description of Revision	Office of Primary Responsibility	Effective Date
1.0		Mr. Daniel Murri, NASA Technical Fellow for Flight Mechanics, LaRC	6/13/13

NESC Request No.: TI-09-00605

Table of Contents

Technical Assessment Report (Phase 2)

1.0 Notification and Authorization 7
2.0 Signature Page 8
3.0 Team List 9
4.0 Executive Summary 10
5.0 Assessment Plan 12
6.0 Problem Description and Proposed Solution 14
7.0 Data Analysis 14
    7.1 AADS Phase 2 Model Improvements 15
    7.2 Improved Simulations 32
    7.3 Monte Carlo Analysis 44
    7.4 Processor Load Analysis/Benchmarking 51
    7.5 Other Trades and Sensitivity Analysis 54
    7.6 AADS Performance Assessment 59
    7.7 Open Phase 2 Investigation Topics 89
8.0 Findings, Observations, and NESC Recommendations 92
    8.1 Findings 92
    8.2 Observations 93
    8.3 NESC Recommendations 93
9.0 Alternate Viewpoint 94
10.0 Other Deliverables 94
11.0 Lessons Learned 94
12.0 NASA Standards and Specifications 94
13.0 Definition of Terms 94
14.0 Acronyms List 95
15.0 References 97

List of Figures

Figure 7.1-1. AADS Phase 2 Block Diagram 17
Figure 7.1-2. Venus Mission Scenario RSE #1 Actual Temperatures versus Predicted Temperatures for Densities between 2 and 30 kg/km^3 23
Figure 7.1-3. Venus Mission Scenario Actual Temperatures versus Predicted Temperatures for Densities between 30 and 60 kg/km^3 24

NASA Engineering and Safety Center Technical Assessment Report	Document #: NESC-RP-09-00605	Version: 1.0
Title: Autonomous Aerobraking (Phase 2)		Page #: 4 of 97

Figure 7.1-4. Venus Mission Scenario RSE #1 Maximum Solar Panel Temperature Residual versus Predicted Maximum Temperature for densities from 2 to 30 kg/km^3 25

Figure 7.1-5. Venus Mission Scenario RSE #2 Maximum Solar Panel Temperature Residual versus Predicted Maximum Temperature for densities from 30 to 60 kg/km^3 25

Figure 7.1-6. Venus Mission Scenario Model Fit Error Distribution.. 26

Figure 7.1-7. Venus Mission Scenario Model Representation Error Distribution 27

Figure 7.1-8. Flowchart of the AADS N-sigma Bias Maneuver Logic 29

Figure 7.1-9. Scaling Flight-like Atmosphere for the ORT Simulation 31

Figure 7.2-1. Reference and Operational Mission Run-outs used to Establish Criteria for Phase 2 AADS Evaluation within an ORT 34

Figure 7.2-2. SP Apses Timing Performance at Mars in AAHFS without Accelerometer Errors.. 37

Figure 7.2-3. SP Periapsis Altitude Performance at Mars in AAHFS without Accelerometer Errors.. 38

Figure 7.2-4. AADS Corridor Performance at Mars in AAHFS without Accelerometer Errors.. 38

Figure 7.2-5. SP Apses Timing Performance at Titan in AAHFS without Accelerometer Errors.. 39

Figure 7.2-6. SP Periapsis Altitude Performance at Titan in AAHFS without Accelerometer Errors.. 40

Figure 7.2-7. AADS Corridor Performance at Titan in AAHFS without Accelerometer Errors, Overlaid by POST2 Results from Phase 1 40

Figure 7.2-8. SP Apses Timing Performance at Venus in AAHFS without Accelerometer Errors.. 41

Figure 7.2-9. SP Periapsis Altitude Performance at Venus in AAHFS without Accelerometer Errors.. 42

Figure 7.2-10. AADS Corridor Performance at Venus in AAHFS without Accelerometer Errors.. 42

Figure 7.2-11. Block Diagram for the EML Version of AADS 44

Figure 7.3-1. AADS Monte Carlos Results: Aerobraking Duration 46

Figure 7.3-2. AADS Monte Carlos Results: Final LTST 47

Figure 7.3-3. AADS Monte Carlo Results: Number of Maneuvers 47

Figure 7.3-4. AADS Monte Carlo Results: Total ABM ΔV 48

Figure 7.3-5. AADS Monte Carlo Heat Rate Distribution 49

Figure 7.3-6. AADS Monte Carlo Results: Heat Rate Statistics 50

Figure 7.3-7. AADS Monte Carlo Results: Periapsis Timing Errors 50

Figure 7.4-1. AADS Low-Rate Integrator State Determination Scaled Time for Mars Case using 21x21 Central Body Harmonic Gravity Model.. 53

Figure 7.4-2. AADS Low-Rate Integrator Orbit Propagator Scaled Time for Mars Case using 21x21 Central Body Harmonic Gravity Model 54

Figure 7.5-1.	Ground Update State Uncertainty Propagated over 7-days	55
Figure 7.5-2.	Corridor Performance for Phase 2 AADS with 7-day Ground Update and Accelerometer Errors	56
Figure 7.5-3.	Apses Timing Performance for Phase 2 AADS with 7-day Ground Update and Accelerometer Errors	57
Figure 7.5-4.	Periapsis Altitude Performance for Phase 2 AADS with 7-day Ground Update and Accelerometer Errors	57
Figure 7.6-1.	Profiles of Observed Density (blue), Model Density with A0 =1 (solid red) and the Model Density Scaled to Match area under Observed Density Profile (dashed red) for a sample AADS Mission Orbit 209	62
Figure 7.6-2.	A0 Multipliers Observed and used in the Simulations	63
Figure 7.6-3.	ORT1 Daily Analysis	65
Figure 7.6-4.	ORT1 Weekly Analysis: Orbit Period	67
Figure 7.6-5.	ORT1 Weekly Analysis: LTST	67
Figure 7.6-6.	ORT1 Heat Rate versus Days of Aerobraking	68
Figure 7.6-7.	ORT2 Heat Rate Corridor	70
Figure 7.6.8.	ORT2 with the 3-Sigma High Heat Rate Predictions	71
Figure 7.6-9.	ORT2a with the Density 3-Sigma Correction	72
Figure 7.6-10.	Data Flow Chart for the Two Types of ORTs; EOM	73
Figure 7.6-11.	ORT Corridors	75
Figure 7.6-12.	ORT Glide Slope Comparison to Reference	76
Figure 7.6-13.	LTST of the Ascending Node Comparison to Reference	77
Figure 7.6-14.	ORT LTST versus Orbit Period Comparison to Reference	77
Figure 7.6-15.	ODY Aerobraking Maximum Heat Rate	83
Figure 7.6-16.	AADS ODY Aerobraking Mission Run-out Maximum Heat Rate	84
Figure 7.6-17.	AADS ODY Narrow Corridor Simulation Peak Heat Rates	85
Figure 7.6-18.	Mission versus AADS Glide Slope Comparison	86
Figure 7.6-19.	Mission Compared to AADS Two Replication Simulations	86
Figure 7.6-20.	MRO Aerobraking Peak Heat Rates	87
Figure 7.6-21.	AADS MRO Aerobraking Mission Run-out Maximum Heat Rate	88
Figure 7.6-22.	MRO versus AADS Glide Slope Comparison	88
Figure 7.6-23.	MRO versus AADS Aerobraking Mission Parameter Comparison	89

List of Tables

Table 7.3-1.	Monte Carlo Parameters	45
Table 7.5-1.	Impact of Gravity Model Errors on SP Accuracy and Timing for 7-day Ground Update at Mars when Compared Against Fixed 21 degree, 21 order Gravity Truth Model	58
Table 7.6-1.	Aerobraking Operations Ground-Based Analysis Teams	60
Table 7.6-2.	ORT Performance Metrics	74

NASA Engineering and Safety Center Technical Assessment Report

Autonomous Aerobraking (Phase 2)

Table 7.6-3.   Issues Related to ORT Simulation and their Impacts and Resolutions 79
Table 7.6-4.   Comparison of ODY Flight and AADS Replicated Missions 83
Table 7.6-5.   Comparison of MRO Flight and AADS Replicated Mission 87

Technical Assessment Report (Phase 2)

1.0 Notification and Authorization

The NASA Engineering and Safety Center (NESC) received a request from Mr. Daniel Murri, NASA Technical Fellow for Flight Mechanics at Langley Research Center (LaRC) to develop an autonomous aerobraking (AA) capability.

The NESC received this request on December 15, 2009. An initial evaluation for all phases of this assessment was approved to proceed at the February 4, 2010, NESC Review Board (NRB). The final report for Phase 1 was approved by the NRB on December 15, 2011. The assessment plan for Phase 2 was approved by the NRB on December 15, 2011.

The key stakeholders for this assessment (all phases) are the NESC (including the technical disciplines of Flight Mechanics; Aerosciences; Passive Thermal; Guidance, Navigation, and Control (GN&C); Software; Loads and Dynamics; Human Factors), and future NASA Programs and Projects that may utilize aerobraking.

2.0 Signature Page

Submitted by:

Team Signature Page on File -8/8/13

_____
Mr. Daniel G. Murri   Date

Significant Contributors:

_____   _____
Ms. Jill L. Prince   Date   Mr. Richard W. Powell   Date

_____
Ms. Alicia M. Dwyer Cianciolo Date

Signatories declare the findings and observations compiled in the report are factually based from data extracted from Program/Project documents, contractor reports, and open literature, and/or generated from independently conducted tests, analysis, and inspections.

3.0 Team List

Name	Discipline	Organization
Core Team		
Dan Murri	NESC Lead, NASA Technical Fellow for Flight Mechanics	LaRC
Richard Powell	NESC Deputy Lead, Flight Mechanics TDT	AMA
Jill Prince	Team Lead	LaRC
Hollis Ambrose	APL Simulation	APL
Angela Bowes	LaRC Simulation	LaRC
David Carrelli	APL Simulation	APL
Alicia Cianciolo	LaRC Simulation	LaRC
John Dec	Thermal	LaRC
Allen Halsell	Aerobraking Analysis/AeroNav	JPL
Chris Johansen	MTSO Program Analyst	LaRC
Robert Maddock	LaRC Simulation	LaRC
Dan O'Shaughnessy	APL Simulation Lead	APL
Fazle Siddique	APL Simulation	APL
Tom Strikwerda	APL Project Manager	APL
Robert Tolson	Atmospheric Modeler	NCSU/NIA
Joseph White	LaRC Simulation	AMA
Tung-Han You	Aerobraking Analysis/AeroNav	JPL
Consultants/Peer Reviewers		
Neil Dennehy	NASA Technical Fellow for GN&C	GSFC
Starr Ginn	Loads and Dynamics TDT	DFRC
Cynthia Null	NASA Technical Fellow for Human Factors	ARC
Steve Rickman	NASA Technical Fellow for Passive Thermal	JSC
Dave Schuster	NASA Technical Fellow for Aerosciences	LaRC
Timothy Brady	Systems Engineering Office	JSC
Steven Gentz	NESC Chief Engineer	MSFC
Administrative Support		
Linda Burgess	Planning and Control Analyst	AMA
Terri Derby	Project Coordinator	AMA
Erin Moran	Technical Writer	AMA

NESC Request No.: TI-09-00605

4.0 Executive Summary

Phase 1 of the Development of Autonomous Aerobraking (AA) Assessment investigated the technical capability of transferring the processes of aerobraking maneuver (ABM) decision-making (currently performed on the ground by an extensive workforce and communicated to the spacecraft via the deep space network (DSN)) to an efficient flight software algorithm onboard the spacecraft. To accomplish this, aerodynamic and thermal models for a representative spacecraft were developed for the onboard algorithm known as AA development software (AADS) and a ground-based "truth" simulation developed for testing purposes. An autonomous ephemeris, atmosphere estimator (AE), and maneuver estimator (ME) were also developed and incorporated into the AADS. Previous aerobraking mission experience [refs. 1 and 2] indicated that an increase in the error of the predicted time of periapsis passage requires frequent (daily) ephemeris updates from the ground using DSN. The Phase 1 simulation analysis of AADS demonstrated the algorithm could provide state estimates within the previously accepted 250-second periapsis timing error for more than a week before requiring a ground update, thus eliminating the need for continuous DSN coverage. AADS was tested in simulation at three destinations, Mars, Venus, and Titan, and included atmospheric perturbations and sensor (inertial measurement unit (IMU)) measurement errors.

This document describes Phase 2 of this study, which was a 12-month effort to improve and rigorously test the AADS developed in Phase 1. Model improvements included rewriting the Phase 1 ephemeris estimator (EE) model, which is called the state propagator (SP) in Phase 2 to more easily accommodate flight software structures and leverage existing flight software architectures. The AE was modified to include the prediction of atmospheric density uncertainty, which was incorporated into the maneuver decision logic. The AADS simulation environment, the Program to Optimize Simulated Trajectories (POST2), was upgraded, decreasing run time and increasing simulation capability. Dispersion analysis, using statistical Monte Carlo approaches and error uncertainties, was performed to more accurately measure the operability of using AA in a perturbed environment. Additionally, initial assessments of the spacecraft dynamic dispersions were evaluated using a 6 degree-of-freedom (6-DOF) simulation. Trade studies were assessed with the SP and introduced the option of using AeroNAV, an onboard orbit determination (OD) tool that incorporates functionality not currently modeled in AADS, such as fault tolerances, orbit safing approaches, and sequence timing information. Studies also provided first order analysis of the onboard computational resources required by the software. Finally, the improved AADS performance was assessed by its ability to reproduce the Mars Odyssey (ODY) and Reconnaissance Orbiter (MRO) aerobraking missions and its performance in an ODY-like aerobraking mission operational readiness test (ORT).

The extensive Phase 2 testing verified that the AADS algorithms were accurate, robust, and ready to be considered for application to future missions that utilize aerobraking and has extended the technology readiness level (TRL) to between five and six. Additional development of AA, including studying the implications of operating the AADS in a real-time environment, and then installing the software onto a spacecraft using AADS during flight in a shadow-mode, where onboard-determined commands would be validated against aerobraking decisions made by the ground staff would further the TRL to eight.

5.0 Assessment Plan

The complete development of the AA capability was planned for four phases. Phase 1, summarized in reference 3, included the development of the AADS and evaluated its performance in a simulated setting at Mars, Venus, and Titan using an MRO like spacecraft. Individual components of the AADS, namely the AE, EE, and ME, were developed and tested in both a performance simulation (e.g., POST2) and a high-fidelity flight-like simulator based on the MErcury Surface, Space ENvironment, GEochemistry and Ranging (MESSENGER) spacecraft simulation. Details of this simulation environment and its applicability to aerobraking are provided in the Phase 1 final report, Section 7.5.2.1. The objective of Phase 1 was to demonstrate the technical feasibility of the software, and the simulation analysis showed that it is feasible for AADS to provide AA control of a spacecraft with ephemeris updates no more than once per week at all three sampled destinations potentially reducing cost incurred by both DSN coverage and ground support.

This report summarizes the activities of Phase 2, which explored more sophisticated schemes for the ephemeris and atmosphere estimation as a means of improving AADS robustness and performance. Improved modeling and error checking were implemented and additional off-nominal stress testing of AADS was performed.

The primary tasks planned for Phase 2 are described below. All primary tasks were accomplished.

- Improve AE and EE
    - Desensitize the AE to EE timing errors and investigate modeling methods independent of altitude. Incorporate 3-sigma estimates of atmospheric density in maneuver estimator logic.
    - Reduce errors in the EE by exploring new integration methods, step sizes, and real-time periapsis timing correction.

- Incorporate operational elements in AADS
    - Streamline code execution.
    - Add error handling and fault detection and correction.
    - Accommodate safe-mode events (safety triggers) and off-nominal scenarios, such as pop-up maneuvers and typical aerobraking spacecraft contingencies.
    - Assess the feasibility of incorporating collision avoidance in AADS.
    - Add flexibility to the AADS maneuver logic and implement versatile control strategies that can meet project specific requirements.

- Improve POST2 and AA High Fidelity Simulation (AAHFS) "truth" models and simulations
    - Complete the integration of Venus and Titan environment models into AAHFS.
    - Incorporate Mars aerobraking flight data as a "truth" atmosphere archive.
- Incorporate uncertainties and perform Monte Carlo analyses of AADS performance
    - Identify uncertainties in "truth" and onboard modeling.
    - Perform Monte Carlo assessment of "truth" simulations.
- Stress-test AADS in POST2 and AAHFS
    - Atmospheric random noise and bias.
    - Initial ground errors.
    - Modeling errors.
    - Fault management logic.
- Conduct trade and sensitivity studies
    - EE functionality versus AeroNav.
    - Effect of data buffer frequency on the EE orbit knowledge.
    - Accelerometer sensitivities (bias, noise, scale factor, alignment, etc.).
    - Analyze corridor width versus maneuver frequency.
    - Evaluate maneuver execution performance sensitivities.
    - Solar radiation pressure.
- Benchmark flight-like processor limitations
- Assess comparison of AADS decisions with mission simulations using ground-based operations teams
    - Run ORT simulating a stressing aerobraking environment.
    - Run ODY spacecraft in AADS using ODY atmospheric data to closely recreate an ODY mission.
    - Run MRO spacecraft in AADS using MRO atmospheric data to closely recreate an MRO mission.

Phase 3 would incorporate the AADS onto a flight-like processor to study the implications of operating the AADS in a real-time environment. This phase would quantify the computational resources required to support AADS in a flight mission and increase the TRL to a six.

In Phase 4, the AADS would be installed onto a spacecraft that will use aerobraking, and then AA would operate during flight in a shadow-mode, where all the steps for AA are performed, but the commands are not executed elevating the TRL to between seven and eight. The onboard-determined commands would be validated against aerobraking decisions made by the ground staff.

6.0 Problem Description and Proposed Solution

To reduce propellant required, NASA uses aerobraking to decelerate an orbiting spacecraft into a desired science orbit around a target planet or moon with an appreciable atmosphere. Detailed descriptions of both conventional and AA are described in reference 3. The problem addressed herein is how to reduce the cost and risk of the conventional aerobraking mission phase by moving much of the decision making onboard the spacecraft. Doing so reduces the cost of the DSN and ground support while reducing risk by allowing precisely targeted maneuvers more often. Initial development of the onboard software is described in reference 3. The primary problem for the Phase 2 effort was to verify, through analysis, that AADS is robust and to demonstrate its capability in flight-like environment.

The proposed solution includes: (1) developing a flight-like environment (i.e., using mission observed aerobraking pass densities) (2) upgrading the AADS algorithms to work in the flight-like environment, (3) including additional dispersions, trade studies and stress testing to verify that AADS can perform adequately in a wide range of simulated conditions. The description of how each facet of the solution was developed and used to verify the robustness and demonstrate the capability of AADS is described in the next section.

7.0 Data Analysis

To demonstrate the robustness of AADS, several model improvements were made since Phase 1. Details of the modifications to the AEs, EEs, and MEs are described in Section 7.1. This section also includes a description of the updates made to the thermal model and the introduction of an alternative EE technique within the AeroNav architecture as an alternative AADS implementation. Section 7.2 summarizes the updates to the POST2 and the AAHFS simulations and additional analysis capabilities utilized by the simulations for Phase 2.

Section 7.3 summarizes the results of the Monte Carlo analysis while Section 7.4 describes the results of processor loads testing and benchmarking. Additional trades to evaluate the robustness of the AADS due to unique parameters are described in Section 7.5. Finally, the overall AADS performance demonstration in a flight-like environment is described and results are presented in

Section 7.6. It is based on these analyses that the recommended next step for AADS is implementation in a flight-like processor to study the implications of operating in a real-time environment.

Section 7.7 lists aspects of the AADS development that were not considered as part of Phase 2 and describe what remains as future work to prepare for Phase 3.

7.1 AADS Phase 2 Model Improvements

7.1.1 Atmosphere Estimator

During the implementation of the Phase 1 AE in AADS, it was found that the AE statistical model was providing uncertainties of the estimated density and scale height that were much smaller than the natural variability seen during previous aerobraking missions. To improve the uncertainty estimation, AE was redesigned to use deviation between the predicted and observed values from the number (N) orbits prior to the one being predicted (i.e., orbit N+1). Previous mission data is not used for prediction as scaled season, altitude, time, and location on the planet does not translate well for accurate prediction. A number of averaging methods were tested against previous aerobraking mission data sets. Methods included arithmetic mean, geometric mean, median and harmonic mean. The goal was to find a method that permitted an unusually large change in density to increase the uncertainty, but not to influence the uncertainty over too long a period of time. For example, if the algorithm averages over the previous seven orbits and then the simulation experienced an orbit with an unusually high density then the method would provide a high mean and standard deviation for the next six predictions. This method is intended to protect against situations that occurred during the ODY aerobraking mission on orbit 106 when the observed heat rate was near the immediate action line after several orbits with considerably lower heat rates.

The second study included selecting an N that is large enough to provide a reasonable set of statistics (mean and standard deviation), but was not so large as to not recognize a change in the variability (e.g., periapsis precessing over the polar vortex). It became clear that different approaches provided the best estimates depending on the latitude of periapsis, but the differences were smaller than the predicted variability and not statistically significant. To maintain simplicity, the arithmetic mean and the standard deviation from the mean averaged over seven orbits (N=7) was selected. This approach was tested during the ORT and performed as well as ground-based human operations teams.

7.1.2 EE using the SP

During Phase 2, the EE algorithm was renamed the SP to more clearly convey its functionality, and was completely redesigned to address several shortcomings of the original Phase 1 design. While the results obtained in Phase 1 served as an adequate proof-of-concept, and the EE demonstrated suitable performance in Phase 1 testing in terms of orbital integration accuracy and execution speed, the software architecture and implementation did not allow ready conversion to a flight environment (i.e., real-time processing). Furthermore, a review of the coding practices and algorithms used for the EE showed that performance improvements (orbital integration accuracy, software execution speed, software memory footprint) were possible with a change in software architecture and modifications to the orbital integration algorithms.

A major change was made to the architecture of the EE algorithm used in Phase 1 to better facilitate use in a real-time environment. Running AADS on a flight processor was not a requirement for Phase 2, but will be necessary in subsequent phases of the project. As such, the software architecture modifications were made to the software during Phase 2 in an effort to save the substantial expense that would come from an architecture change in a later project phase. In Phase 1, an external process, presumed to exist in the host spacecraft flight software, would buffer measurement data for batch processing by AADS. In Phase 2, the SP was designed to run synchronously with the IMU/star tracker output data rate, thereby minimizing the software necessary to use AADS on a host spacecraft. Rather than buffering this measurement data, the host spacecraft need only call AADS at a fixed rate (i.e., 10-Hz). Because each spacecraft may have a different IMU and/or star tracker, these rates are user-selectable, ensuring maximum flexibility of the SP software. The architecture used for AADS in Phase 2 is shown in Figure 7.1-1. In the figure, t, a, q, h and x represent time, acceleration, quaternions, altitude, and state data respectively. The capital letters denote buffered data.

The primary function of the SP is to provide predictions of the 3-DOF spacecraft position and velocity for use in the AE and ME. This requires a high-fidelity integration of the spacecraft translational equations of motion. Ground operations provide the initial condition for the integration (termed a "ground update"), and the spacecraft uses this initial state coupled with onboard force models and measurement data to propagate the ground update forward in time. Because the desired integration interval is long (defined by this project to be no more often than once every seven days), the integration must be performed as accurately as possible because the position errors grow quadratically with time.

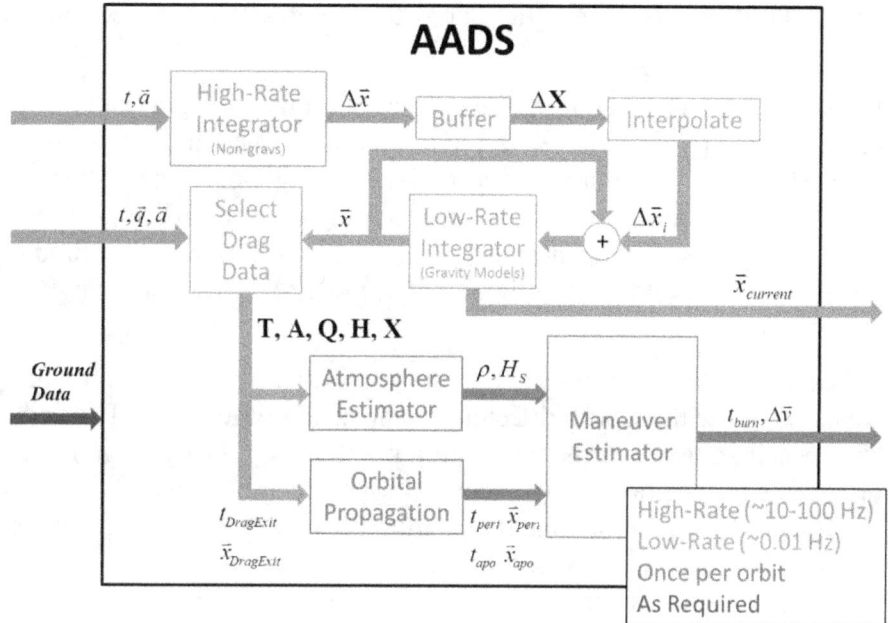

Figure 7.1-1. AADS Phase 2 Block Diagram

The SP is divided into two functions: (1) state determination (SD) or determining the current spacecraft translational state, and (2) orbit propagation (OP) or propagating of the current state forward in time to predict a future orbital condition (periapsis or apoapsis). The SD process integrates the last known state forward in time using the IMU measurements and onboard gravity models to propagate the state up to the current time. Once on each orbit, the OP process takes the current state and predicts ahead one orbit to provide estimates of the subsequent periapsis and apoapsis states to the AE and ME tasks for corridor maneuver computations. The SD runs in real-time, although at a low rate (~0.01-Hz) as it continually updates the translational state based on the sensed accelerations. The OP is a batch process that only runs once per orbit to feed the AE and ME algorithms.

To accurately perform the SD, all accelerations acting on the vehicle must be included in the integration. These accelerations can be broadly grouped into two categories: gravitational and non-gravitational. In general, the gravitational accelerations are modeled in the software, and the non-gravitational accelerations are measured with the accelerometers. The dominant gravitational acceleration is due to the central body, but other perturbing bodies may also be modeled. The dynamics of the motion due to gravity change slowly over time, allowing for a large integration time step. Further, the gravitational accelerations are computationally expensive, so there is a strong desire to limit the number of evaluations of these models. In contrast, the non-gravitational accelerations, which can change quickly with time, particularly aerodynamic drag and thrust acceleration, necessitate a small integration step. The competing step size requirements for the orbital integration create what is known as a computationally stiff system.

The EE solved this stiff system by ignoring the evaluation of the non-gravitational accelerations during the exo-atmospheric portion of the orbit and integrating the equations of motion with a large, variable time step (tens or hundreds of seconds). During those times when the non-gravitational accelerations were substantial (during the drag pass and ΔV maneuvers), the integrator ran with a small, fixed integration step (0.1 second). While this method is reasonably accurate, the high-frequency evaluation of the gravity model(s) during the drag pass and burns is unnecessary and significantly increases computational cost.

The solution to the stiff orbital integration problem used by the SP is to separate the fast and slow dynamics, and integrate them with different methods and time steps. This hybrid method capitalizes on the special structure of the integration problem, $y'' = f(t,y) + g(t)$. The slow dynamics, $f(t,y)$, are due to the gravitational accelerations and are evaluated using a variable-step Runge-Kutta-Nystrom (RKN) method that allows minimal computation of the expensive harmonic gravity models. The fast dynamics of the non-gravitational terms, $g(t)$, are not explicit functions of the state so their contribution to the integrated state can be computed separately and included in the full state integration only when required by the RKN integrator. This high-rate integration of the non-gravitational accelerations is accomplished with a simple fixed-step trapezoidal integration of the input accelerometer data at (or close to) the measurement data rate. This hybrid (fast/slow) integration used in the SP allows rapid fixed-step integration of the fast dynamics, and the variable-step RKN used to integrate the slow dynamics takes steps only as required to maintain the specified integration tolerance. This serves to minimize the computational burden without sacrificing accuracy in the spacecraft SP.

The SP architecture described above and shown in Figure 7.1-1 greatly reduced the AADS memory requirements and computational burden required to accurately integrate the orbit. The largest computational burden in the EE or SP is the harmonic gravity model evaluations. The SP only evaluates these models as necessary (throughout the entire orbit) to maintain tolerance, whereas the EE computes these models dozens of times each second during the drag pass and burns. Depending on the time step of the variable-step integration and the duration of the drag pass, the improved structure of the SP can cut the number of gravity model evaluations by two to ten times over the number required by the EE, greatly speeding the execution of the orbital integration. By integrating the high-rate acceleration data in real-time with a simple integration method, the SP eliminates the need to buffer this data throughout the drag pass. A much smaller buffer (~100 seconds) of high-rate ΔV data is all that is necessary to perform the low-rate integration in the SP versus the large (~4000 seconds) of high-rate accelerometer and quaternion data that was required to execute the EE. This architecture change reduces the memory requirements of AADS by a factor of nearly 100.

7.1.3 EE: AeroNav

AeroNav is a stripped down version of the Jet Propulsion Laboratory (JPL) Navigation AutoNav program. It contains the trajectory integration module with some specific spacecraft dynamic modeling capabilities for autonomous on-board aerobraking/aerosampling applications. AeroNav consists of a RG78 integrator and key spacecraft dynamic modeling functions. It can run onboard the spacecraft and a ground-based simulator. The full AutoNav program has been flown on several spacecraft, including Deep Space 1, Deep Impact, and Stardust nominal and extended missions.

The overall plan for JPL participation in the AA task was to: (1) develop a stand-alone EE for use with AADS and POST2 to serve as a validation point, and (2) feed AADS functions, such as the AE and ME back into the current AutoNav software to produce a functioning onboard system for AA. This would combine the current capabilities for virtual machine language (VML)-based spacecraft commanding of turns and propulsive maneuvers with knowledge of the atmosphere. The final step would be operation of this new AeroNav in testbed mode, with atmospheric modeling based on past operations.

These two tasks proceeded in parallel until the complications of simultaneously modifying and stripping out the EE/SP required more attention. The AutoNav-based EE has been developed and showed good agreement with AADS test cases. The additions to the estimator are described below. To date, the updates to the eventual full AeroNav have been to upgrade to VML 3.0 and to setup a testbed for development.

In future development, incorporating the AE and ME into AutoNav to produce a flight-capable system should be a priority. The lessons learned in Phases 1 and 2 will be used to develop logic algorithms capable of dealing with aerobraking-specific operations, such as corridor control and automated pop-up maneuvers. Running in testbed mode would provide a full verification of AADS.

During AADS Phase 2, several modifications to AutoNav were made to reproduce the EE function. These include low-thrust ion propulsion, executive functions to control the spacecraft (e.g., the commanding of attitude changes or propulsive maneuvers), low-level message-passing functions, optical image processing, orbit-determination filtering, and maneuver-determination functions. The AeroNav delivery retains trajectory integration and the associated spacecraft dynamic modules, drag pass determination and event finding.

High-level function calls are used to issue commands to AeroNav and to return results. Each high-level service is expressed by a C-language function. Each function may have input arguments, and may provide output arguments. Most return a code indicating either success or non success.

AeroNav issues file interface commands to store or retrieve intermediate results in a Unix-like file system (both binary and text files). This has been implemented in order to provide flexibility with multiple languages and environments. It also keeps a number of controlling parameters in text files.

The output spacecraft ephemeris file contains ephemeris polynomials for the spacecraft upon which AeroNav is running. It can be regenerated after each spacecraft event, such as a burn or a drag pass. It can also be re-regenerated after each maneuver computation, to incorporate the upcoming maneuver just prepared. This and other ephemeris files are stored in files of Chebyshev polynomial coefficients. Positions and velocities can be evaluated from these polynomials instead of integrating a trajectory numerically. These binary files have provisions for storing ephemeris coefficients for more than one body. Also, an ephemeris can be segmented time-wise, allowing lower-degree polynomials to be used for different arcs.

The arc for each body contains three sets of polynomials coefficients, for the X-, Y-, Z-position components relative to a specified origin, and with respect to the International Celestial Reference Frame (ICRF)[1]. Polynomials for velocity coefficients are not stored, as they can be derived from the position coefficients.

A number of files in this format are used to provide necessary inputs for the integration. The planetary ephemeris file contains ephemeris polynomials for solar system bodies, such as the Sun, the planets, and their natural satellites. A convention allows storing an ephemeris of the barycenter of a planet/satellite system, as well as ephemerides of the center of each body.

The body parameter file specifies integration center body information, which includes gravity parameters, radius, gravitational constants, and body rotation parameters. It is a text-based input file and must be loaded in the system before calling the ephemeris creation function.

The binary maneuver file contains one record for every maneuver. Each record contains the ET2000 time of the maneuver (from midnight January 1, 2000), and the ΔV components of any maneuver scheduled during the planning cycle. These components are expressed in kilometers/second in the ICRF coordinate system, with the Sun as the origin. The maneuver file should be updated as a maneuver is computed or designed. Note that a finite-burn model is implemented in the latest version, but only an impulsive model is described for this task.

The non-gravitational history file is a binary file describing the non-gravitational forces on the spacecraft (as might be inferred by its name). Each record contains the ET2000 time of the ΔV, and a three-element Cartesian vector of the ΔV components. The vector is expressed in meters/second in the ICRF coordinate frame (not the spacecraft body frame). Typically, accelerometers keep track of thruster firings during deliberate maneuvers, desaturations of the momentum wheels, or drag passes. These are buffered by calls to nongrav_gen.c, which converts high-frequency IMU acceleration data into a lower-frequency non-grav history file.

---

[1] International Celestial Reference Frame; see http://www.iers.org/ (search for icrf)

The user inputs the IMU data, and the desired time step of the output data. The acceleration data are then integrated over each time step and the resultant velocities at the center of each time step are calculated and output into the non-grav file at each time.

A binary OD file contains the ET2000 time and the spacecraft state (position and velocity). This file is used only to store the uploaded OD solution, as AeroNav performs no OD.

Finally, the controls file is intended to hold parameters for all of AeroNav with exception of central body oblateness parameters. Parameters related to the generation of a new spacecraft ephemeris, such as code numbers for the planets/barycenters to include gravitational perturbations, the masses of each planet, and the spacecraft mass and area (only used for solar-radiation pressure calculations) are part of the control file. Also, if desired, there is an overall scale factor for the non-grav history ΔV. An additional parameter specifies the maximum number of non-grav history records to retain. Filenames and directory paths are included in the control file.

A small number of low-level utilities are used to calculate trajectory events; 'dragpass.c' finds the start and end times of the drag pass. The input table of IMU data is searched for the time period during which the acceleration is above a given threshold. The desired threshold is input as a fraction of the peak acceleration, and the duration of the drag pass in seconds is returned.

The 'find_apses.c' searches an input spacecraft Chebyshev ephemeris file for all apses occurring within a specified time period for a specified spacecraft. The Chebyshev polynomials are converted to Cartesian states, allowing RxV to be calculated at each time. This method returns the periapsis or apoapsis, but not the maximum or minimum altitude, consistent with AADS convention. The outputs are the time, radius, and type (apoapsis or periapsis) of each apsis.

7.1.4 Thermal Model Update

The Phase 2 report focuses on AADS algorithm robustness testing and operational performance primarily at Mars due to the availability of aerobraking flight data at the planet. However, at the end of Phase 1, an error was identified in the Venus thermal model [ref. 3]. Below is a description of the approach to and results of the updated Venus thermal model.

Because of its proximity to the Sun, the temperatures due to solar heating at Venus are nearly 4.5 times higher as compared to Mars. Likewise, an aerobraking spacecraft will encounter a higher aerodynamic heating environment. During Phase 1, a sample spacecraft, MRO, along with associated thermal models, was used for the AADS aerobraking assessment at Venus. The thermal model nor the spacecraft were designed to handle the thermal loads at Venus, therefore, a different approach was used in Phase 2.

The original approach utilized a single response surface equation (RSE) for the large day-night temperature (and also density) range at Venus. In Phase 2 it seemed practical to break the density range into separate RSE equations. The first RSE (RSE #1) would cover densities between 2 and 30 kg/km^3 and the RSE #2 between densities of 30 and 60 kg/km^3. The original approach used only a single model, Thermal Desktop®, to develop the RSE. Temperatures were calculated using the Systems Improved Numerical Differencing Analyzer/Fluid Integrator software, which can be invoked from Thermal Desktop®. The new approach combined two models and used the temperatures from Thermal Desktop® in MSC Software Patran™, a method that has heritage from the thermal model development demonstrated on both the ODY and MRO aerobraking missions. Results of the new approach are provided below.

7.1.4.3 Response Surface Goodness of Fit Determination

A face-centered central composite design (CCD) with 13 factors was generated using the JMP® statistical software. The CCD had 26 axial points, 10 center points, and 128 points from the fractional factorial contribution. JMP® automatically reduces the fraction used to compute the fractional factorial contribution as the number of factors increases; in this case the fraction was 1/64th. The temperatures calculated for each of the 164 total runs for both Mars and Venus were analyzed using JMP® where a least-squares-fit was constructed using the stepwise regression option in JMP®. The result of the regression is a quadratic equation, one unique to the Mars mission scenario and one unique to the Venus mission scenario. The coefficient of determination or R^2 adjusted value was measured and used to determine how well the assumed functional form of the response measures the variability of the supplied data. In this case, the R^2 adjusted value measured how well the quadratic response surface represented the variability in the temperatures generated by the design of experiment (DOE) cases. For the Venus mission scenario, the updated R^2 adjusted value for RSE #1 is 0.9952 and for RSE #2 is 0.9993. An R^2 adjusted value greater than 0.9 was desirable, but was not sufficient to determine the goodness of fit of the response surface.

To understand how well the RSE is fitting the response data from the DOE runs, a plot of the actual versus predicted values, a plot of the residual versus predicted values, and the model fit distributions must be examined. The actual versus predicted plot shows the temperatures calculated by the thermal model for the cases described in the DOE plotted against the temperatures calculated by the quadratic RSE and is given in Figure 7.1-2 for the RSE #1 and Figure 7.1-3 for RSE #2. In the figures and throughout the thermal results section of this report, the term "actual" temperatures refers to the temperatures calculated by Patran Thermal® and the "predicted" temperatures are those computed by the RSE. Note that the temperatures for the Venus mission scenario, again based on a Mars spacecraft, are unrealistically high and are therefore scaled to match the maximum temperature calculated for a proposed Venus aerobraking spacecraft; a spacecraft which had a more robust thermal design and had solar panels tailored to minimize the aerodynamic heating.

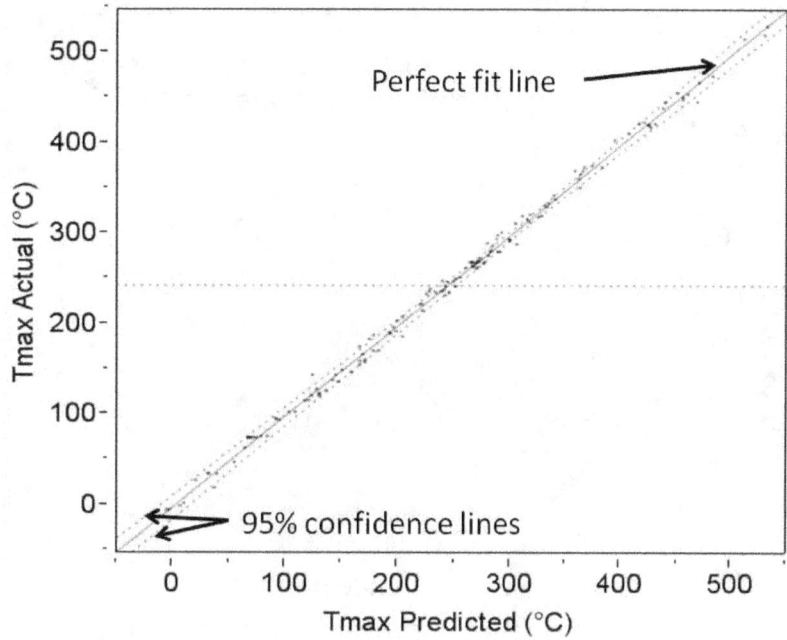

Figure 7.1-2. Venus Mission Scenario RSE #1 Actual Temperatures versus Predicted Temperatures for Densities between 2 and 30 kg/km³

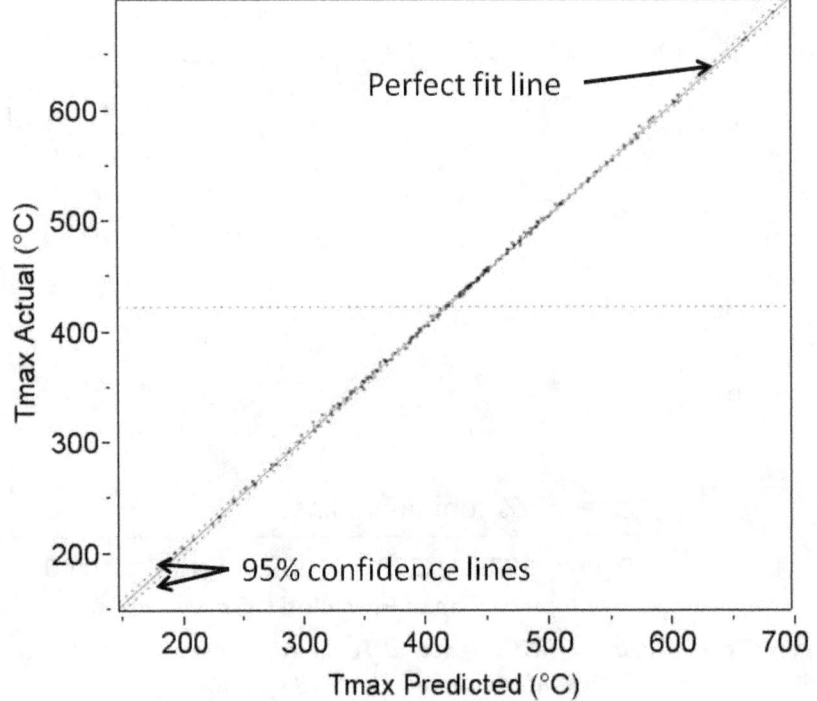

Figure 7.1-3. Venus Mission Scenario Actual Temperatures versus Predicted Temperatures for Densities between 30 and 60 kg/km^3

The solid centerline of the plot in Figure 7.1-2 and Figure 7.1-3 represents a perfect fit of the data. The plots show that the data points lie close to the centerline, which indicates a good fit. The residual is the error between the temperature calculated by the thermal model and the temperature calculated by the RSE. The residual for the maximum solar panel temperature versus the predicted maximum temperature is plotted in Figure 7.1-4 for the RSE #1 and Figure 7.1-5 for RSE #2.

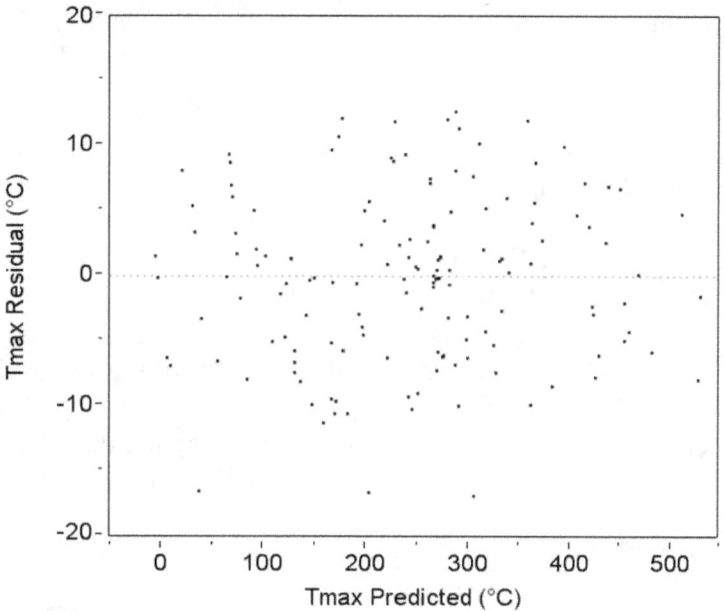

Figure 7.1-4. Venus Mission Scenario RSE #1 Maximum Solar Panel Temperature Residual versus Predicted Maximum Temperature for Densities from 2 to 30 kg/km³

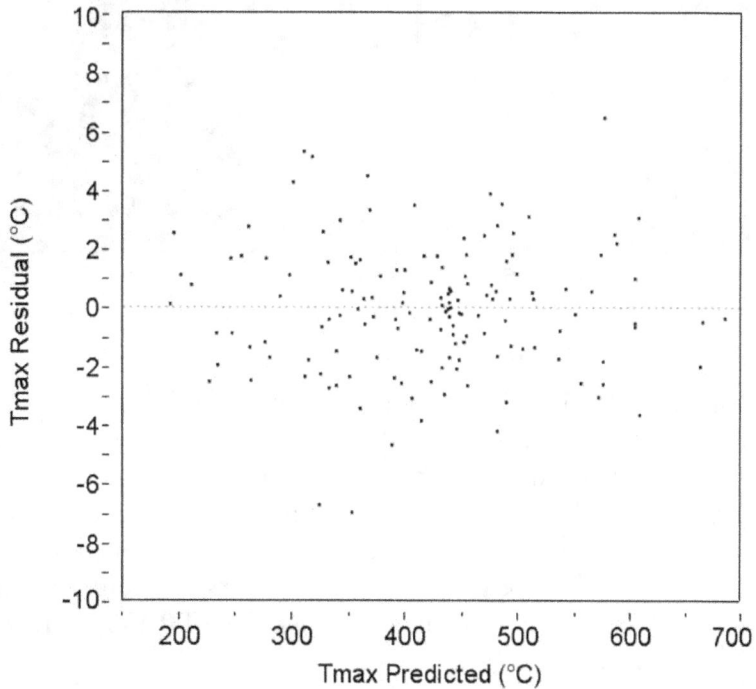

Figure 7.1-5. Venus Mission Scenario RSE #2 Maximum Solar Panel Temperature Residual versus Predicted Maximum Temperature for Densities from 30 to 60 kg/km³

In general, the data points are randomly scattered in Figures 7.1-4 and 7.1-5, which indicates a good fit of the temperature data.

One final check of the goodness of fit is to examine the model fit and model representation error distributions. Both are calculated as percentages based on absolute temperatures in Kelvin. The model error distributions should approximate a normal distribution with a mean around zero and standard deviation less ≤ 1.0. The model fit error is a measure of how well the response surface fits the temperature data in the DOE. The model fit error distribution for the maximum temperature for the RSE #1 and RSE #2 is plotted in Figure 7.1-6. The model fit error distribution for the first Venus RSE is approximately normal and has a mean of -0.0251 and a standard deviation of 1.4326. The model fit error distribution for the second Venus RSE is approximately normal and has a mean of -0.00054 and a standard deviation of 0.3278. The standard deviation for RSE #1 is higher than desirable and may indicate that the density range between 2 and 30 kg/km^3 might be further divided into separate RSEs in future analysis.

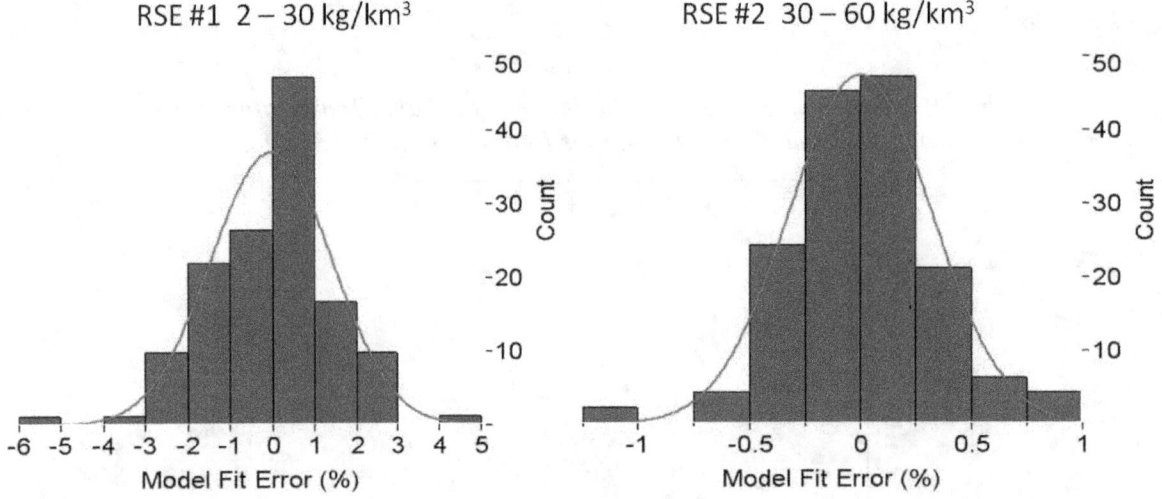

Figure 7.1-6. Venus Mission Scenario Model Fit Error Distribution

The model representation error is how well the response surface fits temperatures calculated by the thermal model as percentage error based on difference in temperature (K) for points other than those on the DOE. The updated model representation error for RSE #1 and #2 for the Venus mission scenario is plotted in Figure 7.1-7. The model representation error distribution for the RSE #1 is approximately normal with a mean of -0.6723 percent and standard deviation of 2.2192 percent. For RSE #2 the model representation error distribution is approximately normal with a mean of -0.0022 percent and standard deviation of 0.4301 percent. Again, RSE #2 appears to have a better fit than RSE #1.

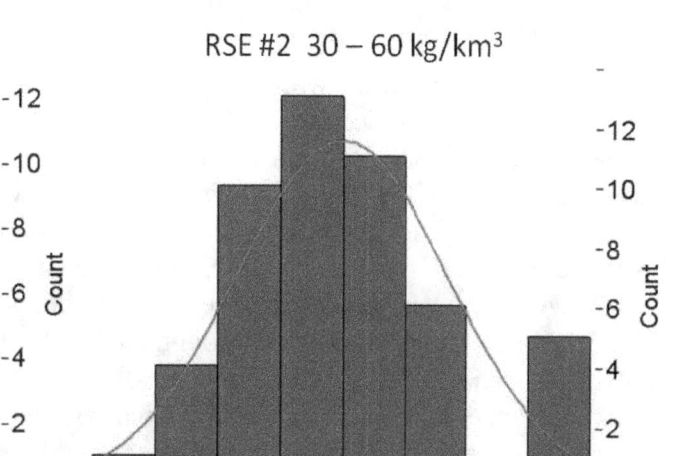

Figure 7.1-7. Venus Mission Scenario Model Representation Error Distribution

The model fit and model representation errors are accounted for in the RSE when the temperature calculation is made from within the AA simulation to provide a conservative temperature. Another error is also added as a bias to the temperature calculated by the response surface. This error is present because the high-fidelity thermal model will typically not be correlated to the aerobraking flight temperature data. This error is typically unknown until the first couple of drag passes are made and the flight temperatures and predicted temperatures compared. Therefore, a short calibration period is required, but this can be accomplished during the walk-in phase of aerobraking, which makes up the first initial orbits where the spacecraft periapsis is gradually lowered into the aerobraking altitude corridor.

One important aspect of response surface modeling that must be emphasized is the RSEs are only valid over the range for which they are defined. It must be stressed that even a small amount of extrapolation in any factor included in the equation can produce results that are invalid.

The Phase 2 analysis indicated thermal model fits could be improved by moving from a single to a double RSE approach. Depending on the specific Venus mission, it may be determined that an aerobraking mission at Venus actually requires several different RSE's to accurately approximate temperatures, however, further analysis and sensitivity studies are needed.

7.1.5 Maneuver Estimator

For the Phase 2 analyses, the ME was updated to utilize the 1-sigma uncertainty on density now provided by the AE (see Section 7.7.1). When the ME predicts the spacecraft's next atmospheric pass maximum heat rate relative to the operational corridor, it also calculates the N-sigma high (or low, in the case of the altitude corridor) estimate, where N is specified by an AADS iLoad parameter. This additional dataset is output for informational purposes, but can also be used to modify the AADS ME maneuver logic. If instructed to do so (through the setting of an additional AADS iLoad parameter), the ME can compare the N-sigma estimate against an "immediate action line" (also input through an AADS iLoad), and if exceeded, the ME passes into its "bias maneuver" logic. The intent of this capability is to further protect the spacecraft from any instances where the AE density uncertainty is sufficiently high as to cause concern that an immediate action line violation has an increased likelihood of occurring on the next atmospheric pass. In this case, the ME will command an altitude raise maneuver such that *at least* the current (previous pass) altitude is maintained. The logic is described below and shown in the flowchart in Figure 7.1-8.

- If the nominal AADS ME logic (looking at only the predicted nominal location with respect to the operational corridor) has indicated an altitude raise maneuver is necessary, where the increase in periapsis altitude places the next periapsis altitude at or above that of the previous pass, that maneuver is commanded; if the altitude raise is insufficient to maintain the previous pass periapsis altitude, the maneuver is adjusted such that the current periapsis altitude is maintained.

- If the nominal AADS ME logic has indicated an altitude lowering maneuver is necessary, this maneuver is ignored, and instead the ME commands only that maneuver which may be necessary to maintain the periapsis altitude estimated from the previous pass; if the next periapsis altitude is estimated to be at or above that of the previous pass without a maneuver, no maneuver is commanded.

- If the nominal AADS ME logic has indicated no maneuver is necessary, the ME commands only that maneuver which may be necessary to maintain the periapsis altitude estimated from the previous pass; if the next periapsis altitude is estimated to be at or above that of the previous pass without a maneuver, no maneuver is commanded.

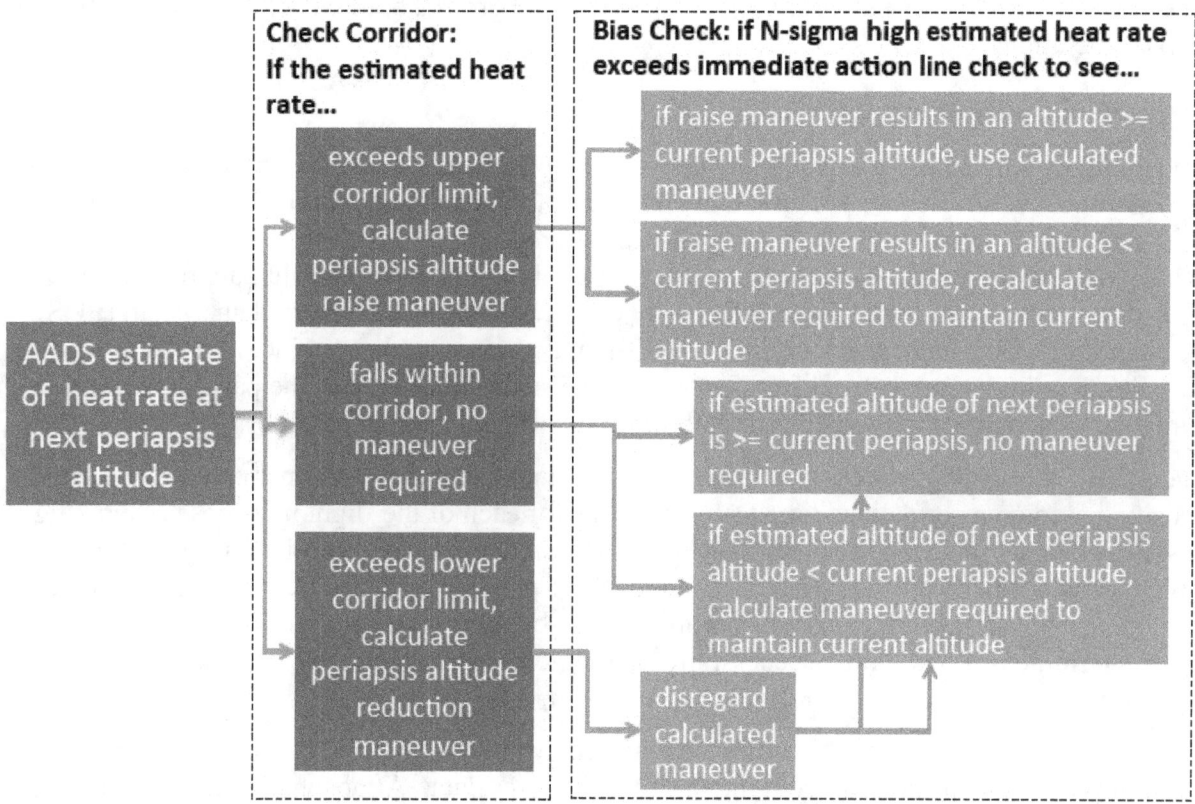

Figure 7.1-8. Flowchart of the AADS N-sigma Bias Maneuver Logic

Pop-up Capability

In addition to the bias maneuver logic, the ME was updated with "pop-up maneuver" logic. In this case, if instructed to do so (through the setting of an additional AADS iLoad), the ME can compare an estimate of the spacecraft's previous pass with respect to the operational corridor against the specified immediate action line, and if the previous pass violated this operational limit, the ME would immediately command a maneuver to raise the spacecraft altitude by a specified (through the setting of an additional AADS iLoad) number of atmospheric scale heights, and AADS operations would cease. In the current implementation, the simulation would simply terminate after the pop-up maneuver is executed, however, operationally, the AADS calculations could be allowed to continue, though any AADS maneuver commands would be ignored by the spacecraft. The intent of this capability is to protect the spacecraft in the event that a likely corridor violation on the previous pass was not predicted during the previous AADS call. The error in the prediction could have resulted from errors in the AADS execution (e.g., AE density predictions and/or SP state predictions) during the previous call, or simply large environmental variations, which were not anticipated. In either case, the spacecraft would be placed into a "safe" orbit and await instructions from the ground before returning to nominal AADS operation.

7.1.6 Atmosphere Modeling

An additional atmosphere modeling update was made for Phase 2 that did not directly affect the AADS code, but allowed its evaluation in a realistic atmosphere, in contrast to the nominal Gaussian density curves generated using Global Reference Atmosphere Models (GRAMs). To evaluate AADS in a flight-like environment, aerobraking atmospheric flight data from Mars Global Surveyor [ref. 6], ODY and MRO were obtained from the planetary data system (PDS). Because the AADS flight trajectory in the ORT simulations would not coincide with the altitude, latitudes, longitudes, season and LTST, nor have the same flight time as the past flight missions, a method to scale the flight profiles in both time and altitude was developed.

The goal of the scaling process is to create a POST2 data table that is a function of altitude and density, based on a normalized time from periapsis, for each of the flight missions aerobraking passes. Then, by indexing the profiles using an input parameter, the simulation can select from any mission's density profile. Therefore, atmosphere profiles from different missions could be used in a single simulation. Likewise, profiles could be used in any order. The objective would be to determine if AADS could successfully accommodate the wide range of observed density structure.

The process to scale the profiles required flight time from periapsis and the corresponding time-dependent altitude and density from the PDS atmospheres data. Atmospheric scale height (e.g., the change in altitude that results in a factor of e change in density), calculated from a fit to periapsis data within 10 to 15 km of periapsis altitude during operations for each profile, is also used in the scaling and is not time dependent.

Flight data is tagged as time from periapsis. Because the flight data time from periapsis is not the same for the inbound and outbound leg of the trajectory, the first step in the scaling process is to normalize time to range from -1 to 1. This is done by taking the shortest leg (either inbound or outbound) of the trajectory and by dividing all times by the lowest absolute value. For example, if a trajectory has a 200 seconds inbound leg and a 150 seconds outbound leg, so that the trajectory time stamps extend from -200 to 150 seconds, all times are divided by 150 seconds, yielding a range of data from 1.33 to 1. Any points outside the range will be ignored in the simulation. Based on the normalized time values, multi-variant POST2 lookup tables based on orbit number and normalized time from periapsis are written for both density and altitude. Scale height tables are also written as a function of orbit number only.

In the simulation an altitude trigger is specified (e.g., 200 km) as the point at which an atmosphere is activated. This is a valid assumption for both long and short period orbits, as there is little appreciable atmosphere above 150 km. When the vehicle is at or below 200 km altitude, the flight atmosphere lookup tables are enabled. Otherwise it is assumed there is no atmosphere.

When the atmosphere is enabled, the time to periapsis and orbit number are used to lookup reference density (ρ_o) and altitude (h_o) values from the tables and scale height (h_s). These values are used, along with current altitude (h), to calculate current density (ρ) in an exponential atmosphere model:

$$\rho = \rho_o e^{\left(\frac{h_o - h}{h_s}\right)}$$

To verify that the actual flight density profiles are implemented in the simulation, the profiles are compared to POST2 profiles generated by the engineering atmosphere model, MarsGRAM. Nominal MarsGRAM aerobraking density passes resemble bell shaped curves indicative of density changing with altitude during an aerobraking pass. Figure 7.1-9 compares orbit densities from MarsGRAM (top) to actual ODY flight data, and ODY flight data scaled to a change in periapsis altitude of -590 m, +1143 m, and +560 m in sample orbits 1, 2, and 3, respectively. The scaling maintains observed density profile characteristics that are intended to stress AADS while accommodating the variations in time and altitude of the ORT simulations. The flight profiles for each mission, MGS and MRO, are scaled in the same manner.

In Phase 2, all three mission data sets were used to evaluate the performance of AADS in realistic flight-like conditions. Results of the evaluation are provided in Section 7.6.1.

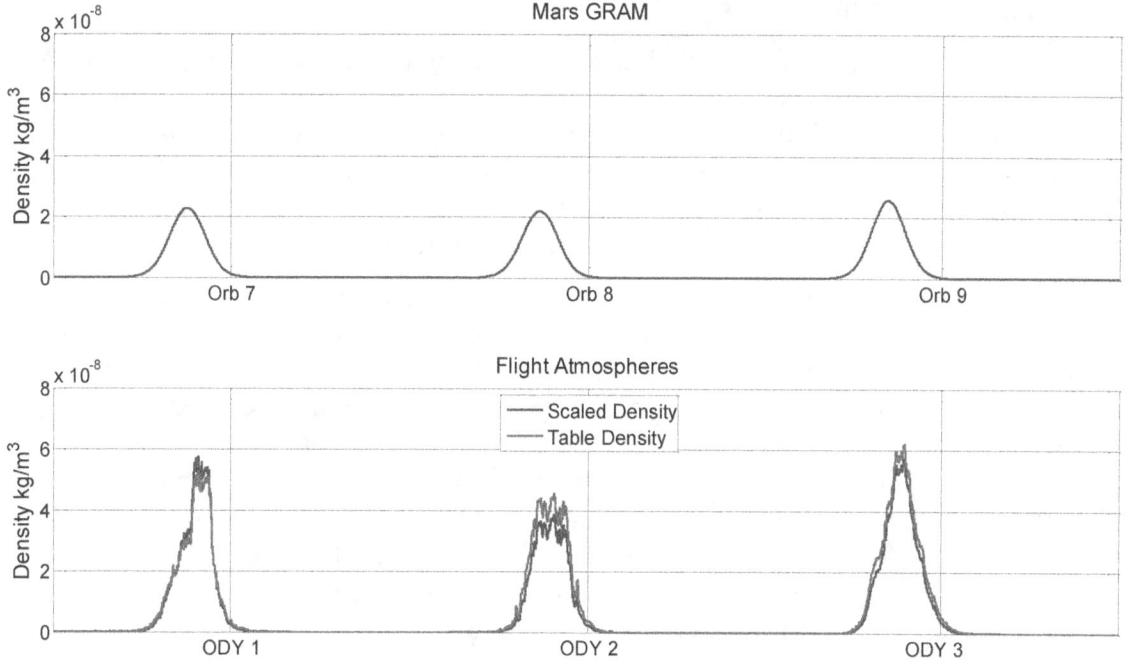

Figure 7.1-9. Scaling Flight-like Atmosphere for the ORT Simulation

7.2 Improved Simulations

Two simulations were used to evaluate AADS during Phase 1: POST2 and the AAHFS. Full descriptions can be found in the Phase 1 report [ref. 3]. The same simulations were used for Phase 2, though several improvements were made. This section describes the simulation improvements that were made and why and how each simulation was used to evaluate the Phase 2 AADS.

7.2.1 POST2

During Phase 2, the POST2 simulation code was updated from version 1.1.8, which was a combined C/Fortran version, to an all-C version. The C version of the performance simulation allows for faster run times than Fortran and allows greater flexibility, efficiency, and capability.

Additional upgrades included adding a Monte Carlo capability and incorporating flight observed atmosphere profiles. These upgrades enabled faster compile and run times and afforded the opportunity to architect the code in such a way that minimized the differences between the AADS and ground-based simulations into a single executable. This greatly simplified efforts to evaluate AADS for the ORT.

Once all upgrades were implemented and verified, the POST2 simulation was used to complete aerobraking mission simulations. However, prior to describing the mission simulations, key aerobraking terms are defined below.

Maneuver Menu: During ODY and MRO aerobraking operations, a preset menu of maneuver sizes was stored onboard the spacecraft to minimize the chance for error when commanding an ABM. The same strategy was used for the Phase 2 AADS ORT analysis. The maneuver menu values are based on those used for ODY and MRO. They are 0.05, 0.1, 0.2, 0.3, 0.5 and 0.6 m/s. The sign of the magnitude was also selected providing capability to both raise and lower the periapsis altitude to generate lower or higher heat rates, respectively.

Corridor: The corridor is defined as the upper and lower limits on the value of the aerobraking parameter. For all AA Mars simulations, the upper limit is maximum heat rate on an aerobraking drag pass. Venus used maximum solar array temperature and Titan used minimum altitude. The lower limit serves to keep the mission in the atmosphere to finish in the specified amount of time while the upper corridor provides desired or required margin to thermal limits of the spacecraft. These values can and do change throughout an aerobraking mission depending on many parameters including, but not limited to, the location of periapsis, the position relative to the glide slope, the desire to maintain spacecraft safety for a specified duration, etc. The corridor values are input parameters set within the iLoads data set for AADS, and they are also set as input parameters to all ground-based aerobraking simulations.

Immediate Action Line: The immediate action line is intended as a fail-safe option. The value of the line is set at a desired margin from the upper corridor. If the spacecraft experiences an unforeseen environment, such as a sudden increase in density that affects peak heat rate, the spacecraft commands a maneuver calculated by AADS at the next apoapsis to raise the periapsis altitude (lowering the heat rate) and protect the spacecraft from further harm.

Aerobraking Margin: Aerobraking Margin: The dominant unknown for aerobraking is the atmospheric density profile. Density is a primary factor for spacecraft survival, as it is a main factor for both structural loads (dynamic pressure) and thermal (heat rate). *The ratio of the density required to reach the immediate action line and the upper corridor is defined as margin.* This density ratio for the Mars aerobraking missions was converted to either equivalent heat rate margin (if thermal is the designing issue – e.g., ODY and MRO), or dynamic pressure (e.g., MGS) for reporting purposes. Note that it is expected that the spacecraft will exceed the upper corridor on many orbits due to the atmospheric uncertainty, but should not exceed the immediate action line. For ODY, the aerobraking margin was 100 percent and was considered an aggressive mission; MRO maintained over 200 percent margin.

During Phase 2, four different types of simulations were used to characterize and evaluate AADS performance and robustness. The application of each of these simulations as they pertain to the AADS assessment is explained in more detail in Section 7.6. Definitions are provided below:

Reference Mission Run-out: An aerobraking mission simulation from walk-in where the spacecraft makes small-predetermined maneuvers to slowly dip periapsis altitudes, within the atmosphere, to a predefined final apoapsis altitude. The reference spacecraft flies through a GRAM atmosphere on each aerobraking pass and a maneuver is allowed at each orbit apoapsis. For the reference mission run-out, each maneuver is targeted such that the aerobraking controlled parameter, in this case, the maximum periapsis heat rate, is targeted to a specified value on every aerobraking pass. For the purposes of this study, the target heat rate value of 0.1125 W/cm^2 was selected such that the final apoapsis altitude of 2800 km was achieved after approximately 70 days of aerobraking. This mission simulation *provides the reference glide slope, or an apoapsis vs. time curve, that is used as a target value by the other simulations that use AADS.* A reference mission run-out is a standard practice for all aerobraking missions and provides a guideline (glideslope) by which real-time analyses can be compared.

Operational Mission Run-out: A mission run-out is a simulation of a spacecraft through the aerobraking phase that has applied realistic constraints as traditional aerobraking ground simulations. Historically, during ground-based aerobraking operations, maneuvers at apoapsis are limited to once during a specified time increment (e.g., every 48 or 72 hours) to minimize staff workload. The operational mission run-out simulation (similar to the reference mission run-out simulation in that it flies the reference spacecraft through MarsGRAM atmospheres from walk-in through a final apoapsis altitude of 400 km) only allows a maneuver on specified apoapses and imposes limits on the aerobraking parameter (e.g., heat rate). The heat rate limits, also called the corridor, have two main purposes: (1) to maintain spacecraft safety and (2) to control the length of the mission. Further explanation of the corridor is provided later in the section. At an apoapsis designated for a maneuver, the simulation predicts ahead (propagates the orbit) for the specified time to determine if, on any pass through the atmosphere, the heat rate limits are violated. For this study, the heat rate limits were adjusted so that the reference and operational missions end at approximately the same time. Therefore, *the purpose of the operational mission run-out is to obtain an initial guess for the corridor limits that will be used for the ORT.* Additionally, maneuvers are only made when a predicted maximum heat rate is outside the corridor limit. The maneuver size is not optimized to provide the exact target heat rate (e.g., 0.1125 W/cm^2), rather it calculates the maneuver size required to reduce the predicted orbit's maximum heat rate back to the middle of the corridor. A detailed description of the corridor as it pertains to aerobraking is provided in the section below.

Figure 7.2-1 shows the maximum heat rates near periapsis for the reference and operational mission run-outs. The operational mission run-out determined that the initial lower and upper corridors set at 0.85 and 0.14 W/cm^2 respectively, would provide an approximate operational mission ending in nearly 70 days. The immediate action line was defined as 100 percent of the upper corridor or 0.28 W/cm^2 to provide margin similar to the Mars ODY aerobraking mission.

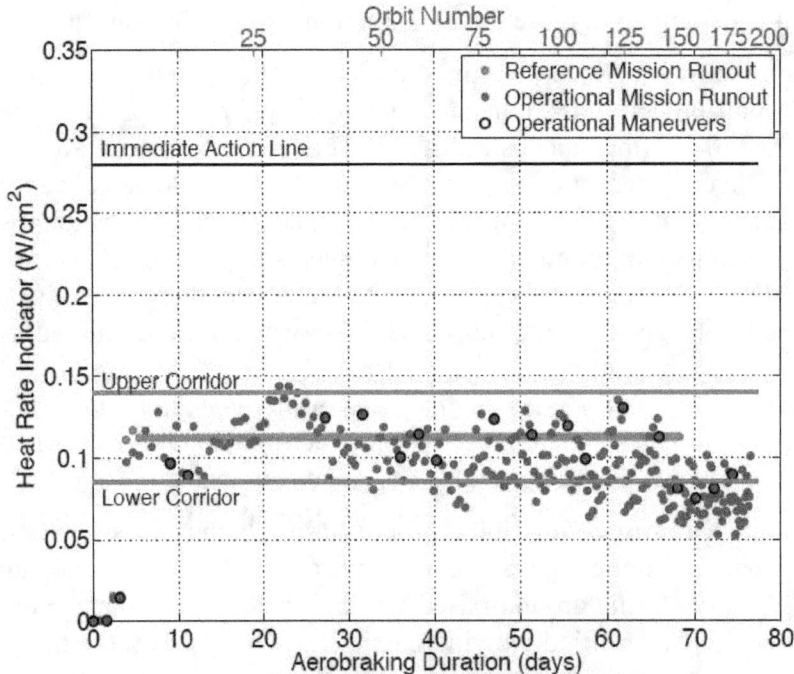

Figure 7.2-1. Reference and Operational Mission Run-outs used to Establish Criteria for Phase 2 AADS Evaluation within an ORT

ABM Sweep: This simulation considers only a short subset of the whole aerobraking mission, known as the aerobraking planning period (APP). The APP is the specified future time span (e.g., up to 1 week) for which a maneuver decision needs to be made and the subsequent passes through the atmosphere are evaluated to verify there are no violations of the corridor. The simulation contains a menu of ΔV options available at the next available apoapsis opportunity (see Maneuver Menu below). The simulation runs a separate trajectory for each option, performing the maneuver at the specified apoapsis and allowing the orbits to propagate through subsequent aerobraking passes until the specified time is met. The resulting max heat rates from each simulation are plotted and evaluated. A ground-based maneuver recommendation is made based on the simulated maneuver trajectory. The decision process is outlined in more detail in Section 7.6.1.3. *This simulation provides analysis with which to base the maneuver recommendation.*

Weekly Reset: The weekly reset simulation is similar to the reference mission run-out in that it allows a maneuver on every apoapsis and targets a specified heat rate. However, the simulation is initiated multiple times with the latest spacecraft state and simulates flight assuming the spacecraft flew exactly the upper, middle and lower corridor heat rates. This simulation *offers an assessment of the end game performance margins.*

AADS Mission Run-out: The AADS mission run-out is similar to the reference mission run-out, but it differs in how and when a maneuver is performed. The AADS uses its prediction of the atmosphere and the SP to determine on which apoapsis to perform a maneuver and how big the maneuver will be. The AADS mission run-out can allow a maneuver on every orbit.

Spacecraft Flight Simulation: Also known as the "truth" simulation, this simulation emulates the "real" environment and the spacecraft flight though it. The simulation uses the same spacecraft parameters, but flies through observed Mars atmosphere aerobraking density profiles rather than MarsGRAM. The simulation is intended to emulate the NAV function of uploading a maneuver to the spacecraft and the DSN communication link providing the latest spacecraft state to the AADS and ground-based simulations.

7.2.2 AA High-Fidelity Simulation

The bulk of the truth model development for the AAHFS was done in Phase 1. This simulation environment is capable of 6-DOF simulations for aerobraking missions at Mars, Venus or Titan. In Phase 1, each of these central bodies used the GRAMs (MarsGRAM, VenusGRAM, and TitanGRAM), but Phase 2 added the capability to run simulations by using the nondimensionalized flight atmospheric data described in Section 7.1.6. This allows 6-DOF testing of the AADS with atmosphere data that reflects characteristics that have been observed by prior aerobraking missions (ODY, MRO and MGS). Although the AAHFS was not used as a part of the ORT conducted in late January 2013, this capability was put in place to allow 6-DOF testing of ORT-like mission scenarios. Test results using these flight atmospheres are provided in the next section.

7.2.2.1 AAHFS AADS Testing

In Phase 1, testing in AAHFS was performed for Mars aerobraking scenarios. Following the development of the SP in Phase 2, AADS was tested in AAHFS for Mars, Venus and Titan. All bodies were tested using 6-DOF simulations, although the attitude dynamics were simplified by reducing the magnitude of the aerodynamic torques. This simplification was made to accelerate the development of the test cases since a substantial effort would be required to modify the vehicle aerodynamic characteristics and attitude control system design to produce well-behaved rotational motion during (and following) the drag pass. This control system design effort for the notional vehicle used for AADS testing was deemed beyond the scope of this project, and this design would follow established procedures and is not considered germane to demonstrating an AA capability. Any mission using AADS would use the detailed vehicle and attitude control system design, and the AAHFS could easily be adapted to these characteristics.

AAHFS –AADS at Mars

Initial testing in a simulated Mars environment highlights the performance advantages of the SP used in Phase 2 over the EE used in Phase 1. Figure 7.2-2 shows the performance of the SP at Mars; the figure shows errors in the AADS translational state when compared with the truth model orbital integration. The SP is able to accurately maintain the current spacecraft state via the state determination (SD) process, as the reconstructed periapsis time drifts by about 0.3 second in the worst-case over a 28-day integration arc. (The term "reconstructed" is used in this context to mean a state prediction that makes use of historical accelerometer measurements, and this state prediction is at the current time. This is opposed to a predicted state, which may be as much as one orbit into the future.) This shows that given near-perfect acceleration measurements by the IMU, the onboard integration can be nearly as accurate as one done on the ground. This is markedly (~10-30 times) better than the performance demonstrated by the EE in Phase 1 within this same simulation. The predicted timing (generated by the OP) for periapsis and apoapsis show similar accuracy. The spikes in the periapsis timing prediction occur when the spacecraft executes a corridor control maneuver, as the OP has no knowledge of the maneuver in the orbit prediction process. Therefore, the apoapsis maneuver introduces a perturbation into the predicted periapsis timing, but it does not affect the apoapsis timing as the maneuver immediately follows apoapsis. Figure 7.2-3 shows a similar trend for the reconstructed periapsis altitude, indicating that the altitude error is less than 40 m after 28 days of propagation. This accuracy is well below that required to perform the AE and ME tasks, even with ground updates 28-days apart (four times the study-defined requirement of weekly ground updates). Of course, the accuracy in the integration is driven by errors in the accelerometer measurements, which are not included in this simulation, but this case does demonstrate excellent accuracy of the SP. Using the SP does not markedly improve the corridor performance, as shown in Figure 7.2-4, as the uncertainty in the corridor control variable (e.g., heat rate) is dominated by the atmospheric variability and not driven by errors in the periapsis altitude or velocity predictions. However, while the error introduced by atmospheric variability does not appreciably grow with time, the state prediction error grows quadratically with the integration interval, so increasing intervals between ground updates can lead these SP error terms to have a significant effect on corridor performance.

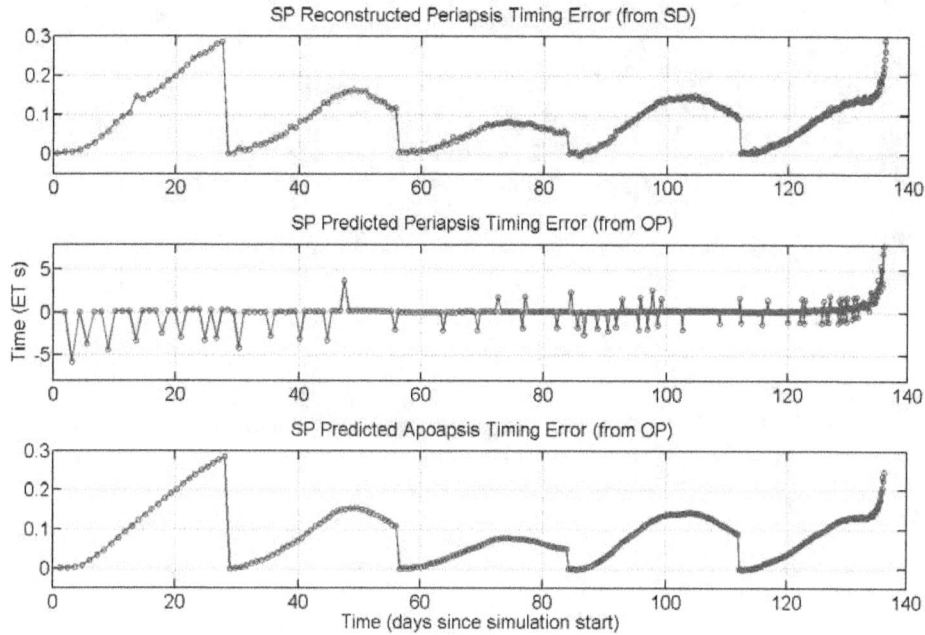

Figure 7.2-2. SP Apses Timing Performance at Mars in AAHFS without Accelerometer Errors

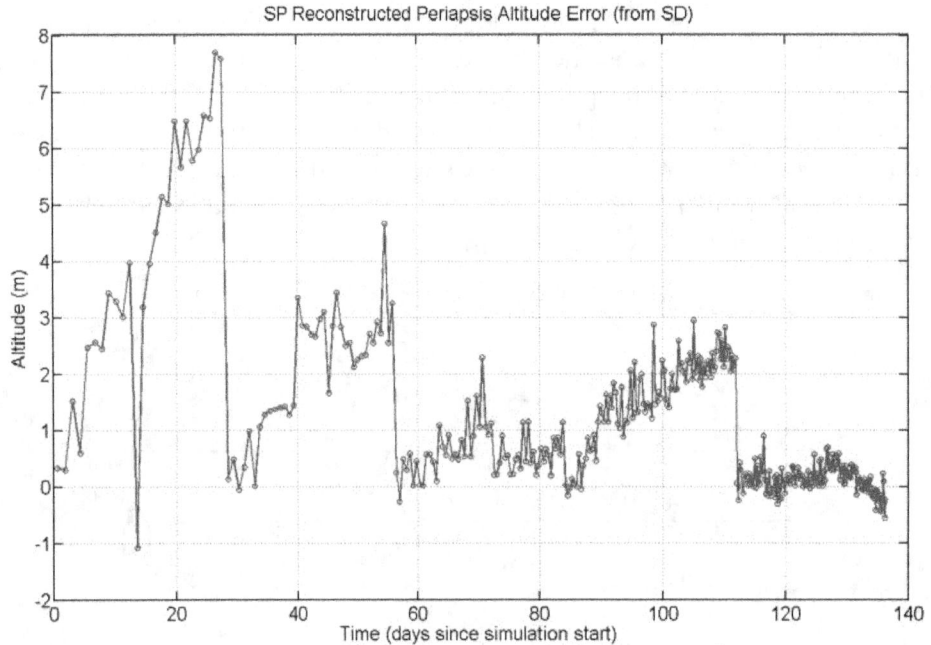

Figure 7.2-3. SP Periapsis Altitude Performance at Mars in AAHFS without Accelerometer Errors

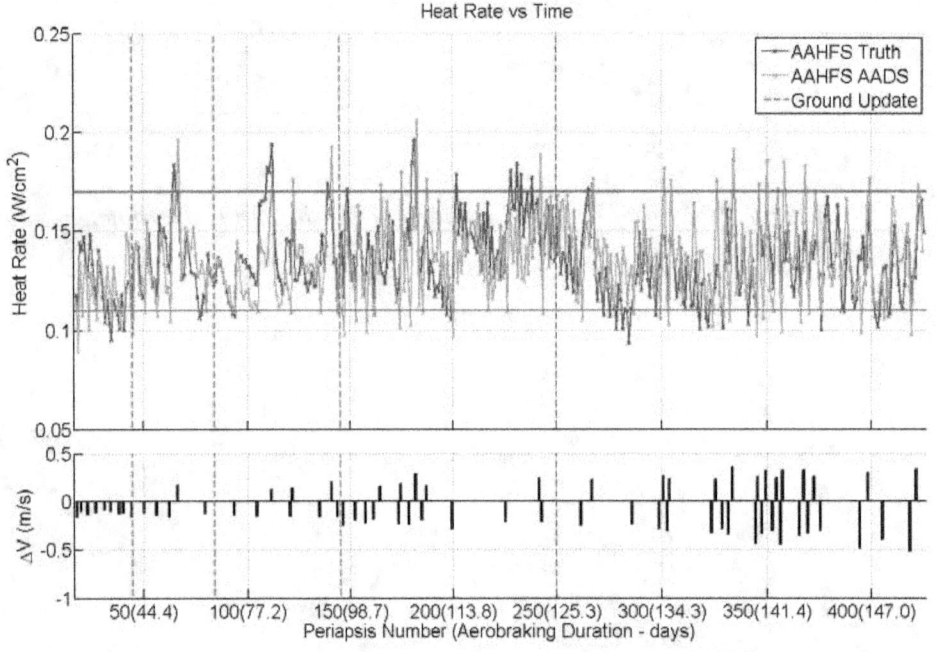

Figure 7.2-4. AADS Corridor Performance at Mars in AAHFS without Accelerometer Errors

AAHFS –AADS at Titan

AAHFS testing was conducted with a mission scenario at Titan, and these results are shown in Figures 7.2-5 through 7.2-7. As with Mars, the SP is able to accurately propagate an initial state over a ground update interval of 28 days. Apses timing errors remain small during this 28-day integration interval; although, the larger maneuvers necessary at Titan (a factor of ~50 larger than those at Mars) introduce a correspondingly larger spike in the periapsis prediction accuracy from the OP. Periapsis altitude errors remain below 70 m worst-case, which is negligible when compared with the 150 km altitude corridor used in the simulation. This corridor is shown in Figure 7.2-7, and the AAHFS results are nearly identical to the POST2 results shown in the Phase 1 final report. The Titan atmosphere does not show the same variability as at Mars, and the smooth nature of the corridor plot reflects this atmospheric uniformity.

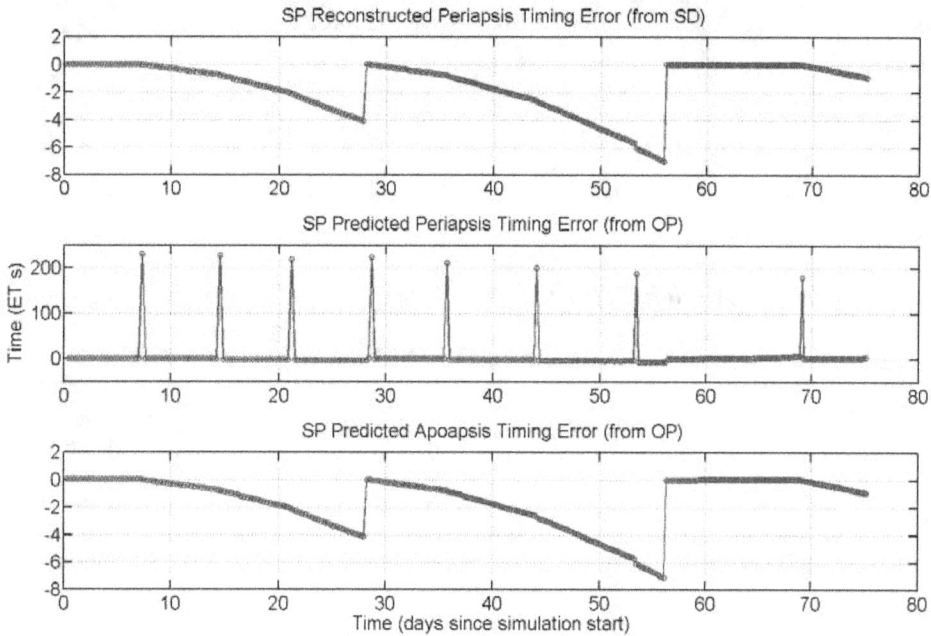

Figure 7.2-5. SP Apses Timing Performance at Titan in AAHFS without Accelerometer Errors

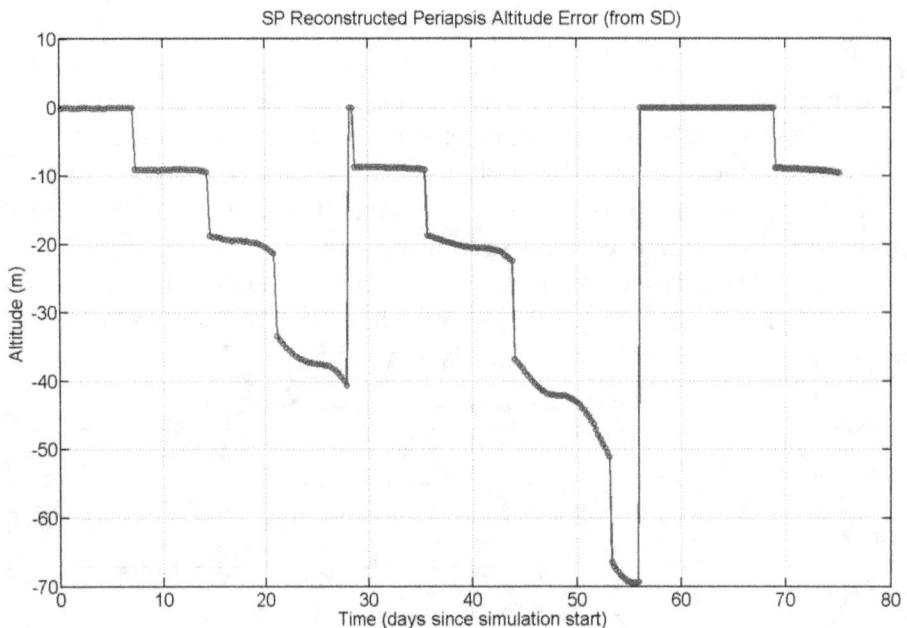

Figure 7.2-6. SP Periapsis Altitude Performance at Titan in AAHFS without Accelerometer Errors

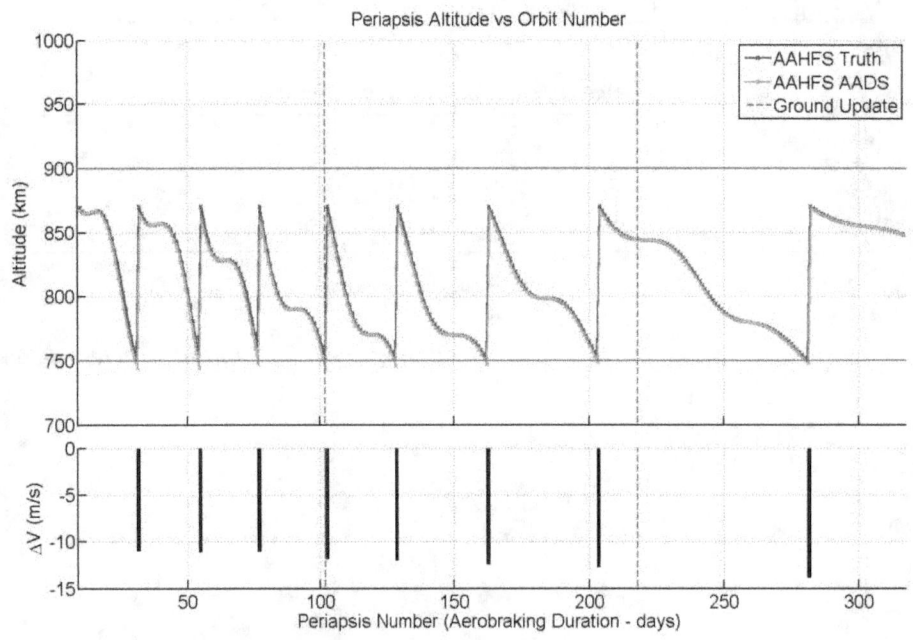

Figure 7.2-7. AADS Corridor Performance at Titan in AAHFS without Accelerometer Errors, Overlaid by POST2 Results from Phase 1

AAHFS –AADS at Venus

AAHFS testing was conducted with a mission scenario at Venus, and these results are shown in Figures 7.2-8 through 7.2-10. As with Mars and Titan, the SP is able to accurately propagate an initial state over a ground update interval of 28 days. Apses timing errors remain small during this 28-day integration interval. Periapsis altitude errors remain below 15 m worst-case. Venus simulations use a temperature corridor, exercising the thermal RSE software inside the AADS. For this case, there are no truth model results shown in Figure 7.2-10, as there is no thermal truth model running in AAHFS. For this case, it is useful to compare the AAHFS results to the POST2 cases shown in the Phase 1 final report; such a comparison shows that the AADS performance remains consistent between AAHFS and POST2, as expected.

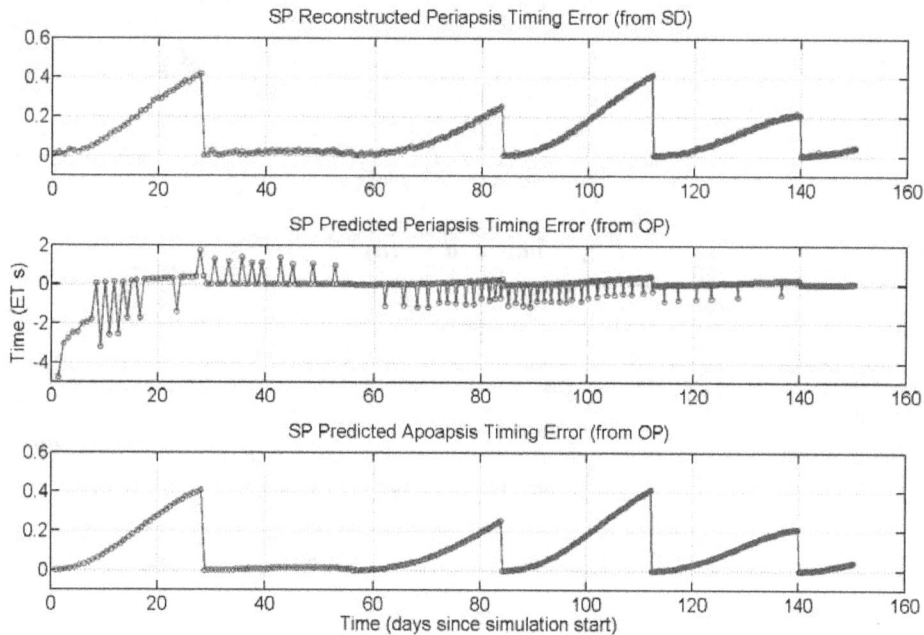

Figure 7.2-8. SP Apses Timing Performance at Venus in AAHFS without Accelerometer Errors

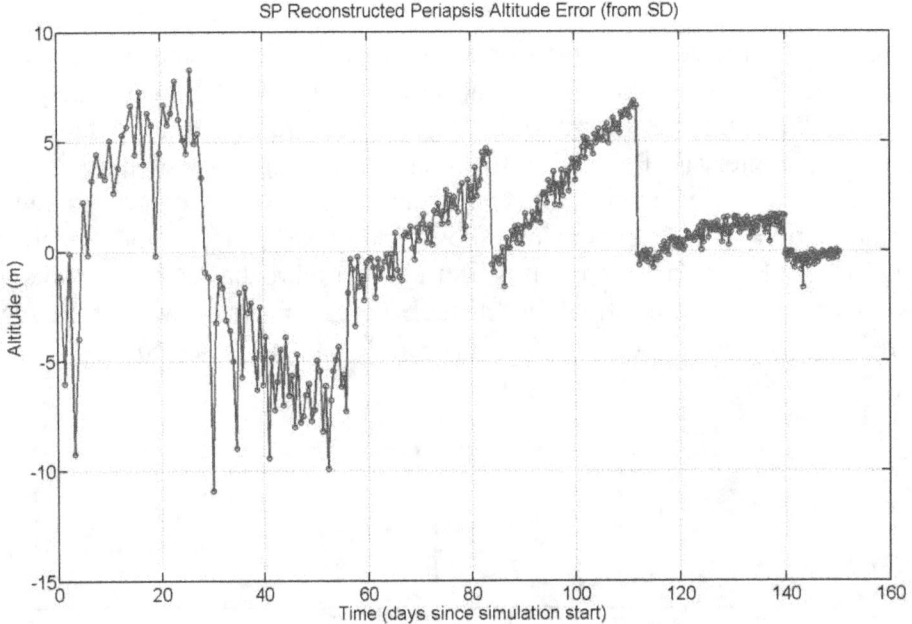

Figure 7.2-9. SP Periapsis Altitude Performance at Venus in AAHFS without Accelerometer Errors

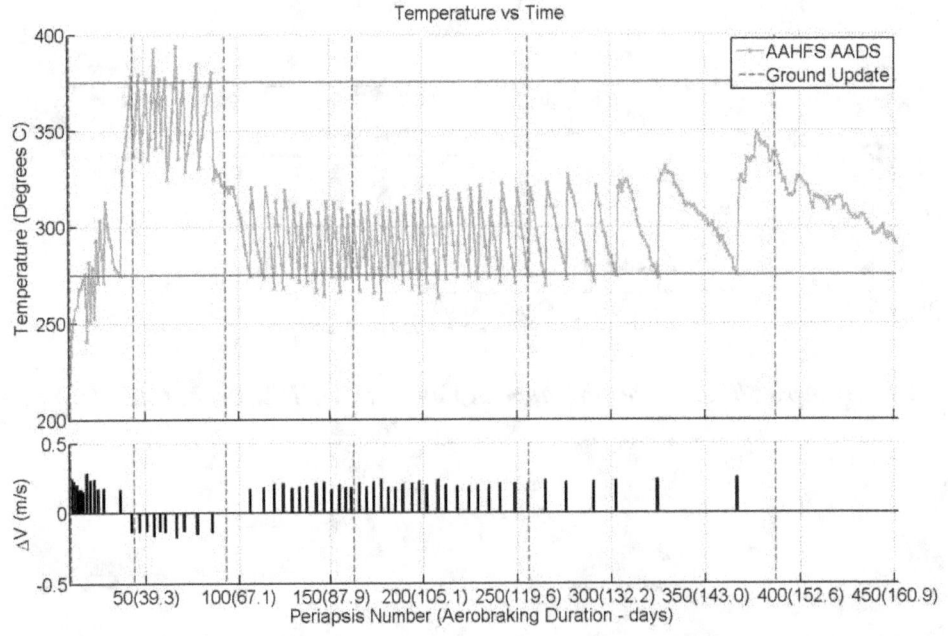

Figure 7.2-10. AADS Corridor Performance at Venus in AAHFS without Accelerometer Errors

The results presented in this section demonstrate consistent performance for AADS when comparing results from AAHFS with POST2 for a variety of conditions. The POST2 results are shown in the Phase 1 final report. Three different central bodies were used (Mars, Titan, and Venus), as well as three different corridor types (heat rate, altitude, and temperature). These results also demonstrate that when the SP provides nearly perfect accelerometer data, the AADS orbit propagation can retain sufficient accuracy over long time spans (28 days or longer) to have a negligible contribution to corridor control errors. Note that the simulated accelerometer data does not perfectly represent the true acceleration as software latencies and quantization error are an inevitable part of the IMU modeling, and these data in the above simulation results do not contain noise, bias, alignment or scale factor errors which be present in flight accelerometers.

7.2.2.2 Embedded MATLAB (EML) Version of AADS

With the proof-of-concept Phase 1 design complete, a more holistic approach was possible to reduce software redundancy and streamline the algorithms. The functional decomposition of AADS in Phase 1 allowed different institutions to bring their expertise to bear on the segments of the AA problem, but with a completed software design, streamlining opportunities were apparent. In order to realize additional performance gains, an effort was undertaken to convert all AADS code to EML. This process allowed restructuring algorithm boundaries, allowing further reductions of the code memory footprint and computational overhead. The block diagram for this implementation is shown in Figure 7.2-11. With the additional algorithm modifications to the AE made in Phase 2, the terms in the regressor and observation matrices in the least-squares estimation problem (A_{LS} and b_{LS}) can be built recursively, eliminating the need to batch process a large vector of input data in the AE, and thereby obviating the need to pass any buffered data to the AE. Although not readily apparent in the block diagram, by converting to EML, the software tasks are easily grouped by execution rate, making this version of the software more amenable to adaptation for an embedded environment. It also streamlines the software interfaces and data structures, eliminating redundancy and making more consistent, readable code. EML also offers rapid, near push-button C-code and executable generation, suitable for a variety of host environments, affording developers the ease of the MATLAB coding and debugging environment and the simulation performance of compiled code. Ultimately, this makes the code more portable and adaptable for use in a variety of flight mission host spacecraft.

Because this EML version of AADS was constructed late in Phase 2, this code was only tested in the AAHFS environment, and not in POST2. Plots for the EML results for Mars, Venus, and Titan are qualitatively identical to the C-version of AADS, and are therefore not shown. Minor algorithm modifications described above prevent quantitatively identical results, as expected. Benchmarking the execution speed of the EML AADS against the C-version should be performed in further development of AA.

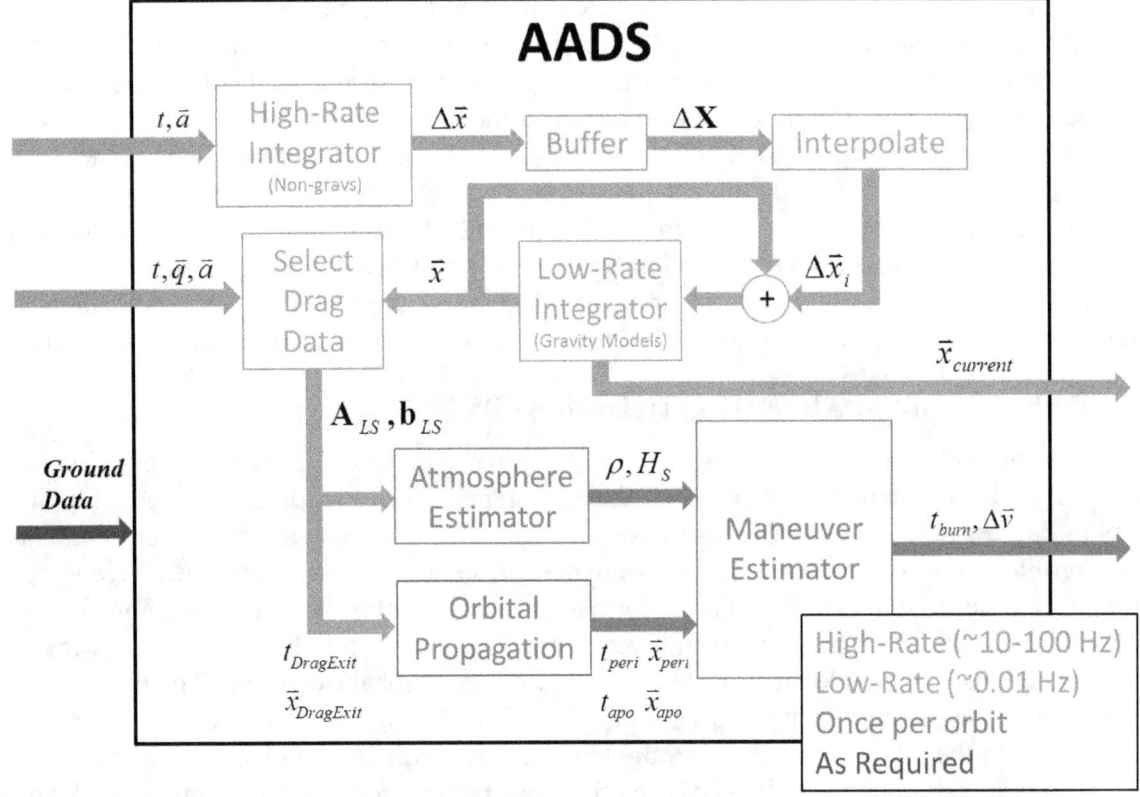

Figure 7.2-11. Block Diagram for the EML Version of AADS

7.3 Monte Carlo Analysis

The objective of the AADS Monte Carlo analysis was to demonstrate that AADS is robust to environmental dispersions like maneuver errors, and aerodynamic and atmosphere uncertainties. The analysis was performed using 2000 3-DOF POST2 simulations of the AADS mission run-out. It should be noted that there was no Monte Carlo of the AADS module (i.e., no multiple simulations on the spacecraft computer). Each AADS mission run-out simulation was evaluated to determine AADS performance degradation as compared to a nominal AADS mission run-out. The full list of Monte Carlo parameters considered is shown in Table 7.3-1.

Table 7.3-1. Monte Carlo Parameters

Parameter	Nominal	Perturbation	Distribution	Rationale
Maneuver Pointing Errors	Inertial α, β, γ	+/- 5 deg	Normal	Angular offset applied to apoapsis velocity vector.
Maneuver timing errors	0.0	+/- 100 s	Normal	Estimated apoapsis time +/- perturbation to see effect of starting burn early/late.
ΔV (m/s) / Maneuver Burn Duration	AADS calculated ΔV	+/- 5 %	Normal	Normally, engine is "on" until AADS calculated ΔV is achieved. To handle burn duration and ΔV parameters, allow applied ΔV= AADS calculated ΔV + perturbation.
Lift Coefficient Multiplier	1	+/- 10 %	Normal	Larger than expected multiplier provides sensitivity analysis.
Drag Coefficient Multiplier	1	+/- 10 %	Normal	Larger than expected multiplier provides sensitivity analysis.
Atmosphere Random Seed	1	1-29999	Integer Uniform	Value used by the atmosphere program to determine perturbation of the density and wind profiles. The range that is allowed by the Mars atmosphere program used in the simulation.
Dusttau (Mars only)	0.5	0.1:0.9	Uniform	This determines the dust loading and thus the density and wind profiles that the vehicle will experience. This range provides large variability, but would not include dust storms.

Histogram results of the primary output parameters of interest are provided. Note that all of the Monte Carlo input parameters are varied normally except for those related to the atmosphere; however, the distribution in the output parameters was not expected to be Gaussian due to the fact that many of the processes involved in aerobraking (e.g., utilizing a corridor) are non-Gaussian in nature. Monte Carlo analyses were performed to evaluate the effect of the corridor on the distributions by flying a narrow corridor at the target heat rate value. The results verified the corridor was a primary driver of the atypical distributions. Further analysis could also be performed to determine the extent of atmosphere effects.

Results of the AADS Monte Carlo analysis indicate that AADS is robust to typical spacecraft aerodynamic, spacecraft maneuver, and atmospheric variations experienced at Mars. The final aerobraking duration for all 2000 cases fell within +/- 7 days of the nominal AADS mission run-out aerobraking duration. The statistics are shown in Figure 7.3-1. The green dot in the histograms denotes nominal AADS mission run-out value while the red dot denotes the mean of the 2000 Monte Carlo simulation.

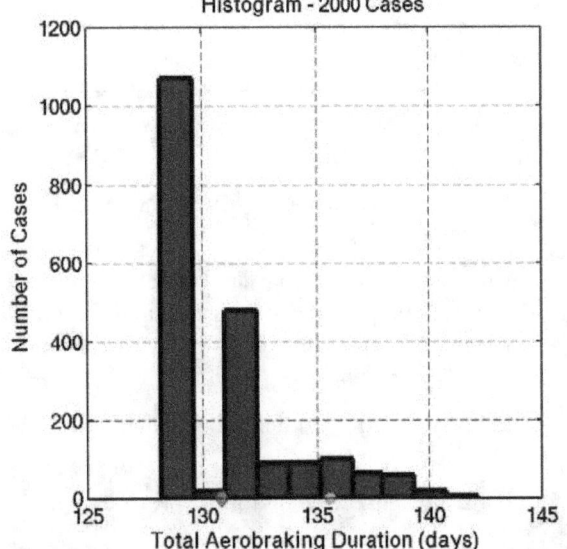

Figure 7.3-1. AADS Monte Carlos Results: Aerobraking Duration

The variation in the local true solar time (LTST) of the equatorial ascending node crossing at the end of each Monte Carlo AADS mission run-out was less than 15 minutes (3-sigma) from the nominal, see Figure 7.3-2. This is consistent with the actual performance criteria imposed on the ODY and MRO aerobraking missions for LTST at the end of aerobraking. Note that AADS meets the performance criteria used in previous missions without any intervention once it is started during the aerobraking phase of the mission. In actual flight operations, the corridor could always be adjusted in a weekly reset meeting such that AADS could end closer to the mission desired LTST if the performance criteria for another mission is less than 15 minutes.

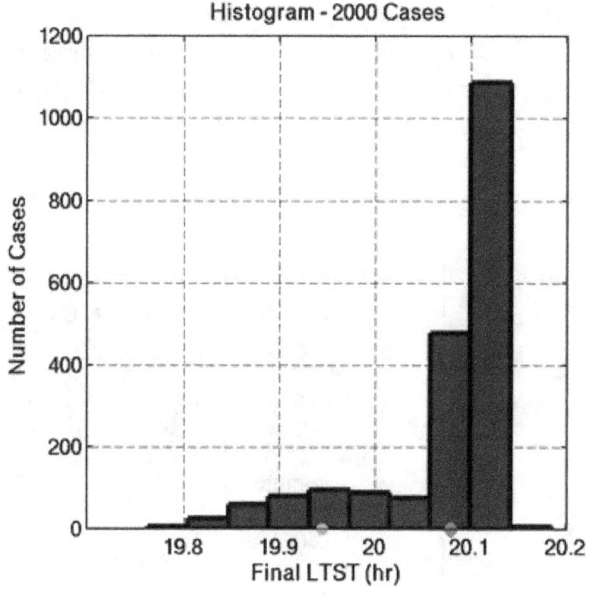

Figure 7.3-2. AADS Monte Carlos Results: Final LTST

The 3-sigma variation in the number of AADS commanded maneuvers, required over the duration of each AADS mission run-out Monte Carlo case, was approximately 13, with the maximum case at 80 (versus 46 for the nominal AADS mission run-out). This variation is quite small given the nominal AADS mission run-out required ~425 orbits, and AADS can perform a maneuver each orbit apoapsis if necessary, see Figure 7.3-3.

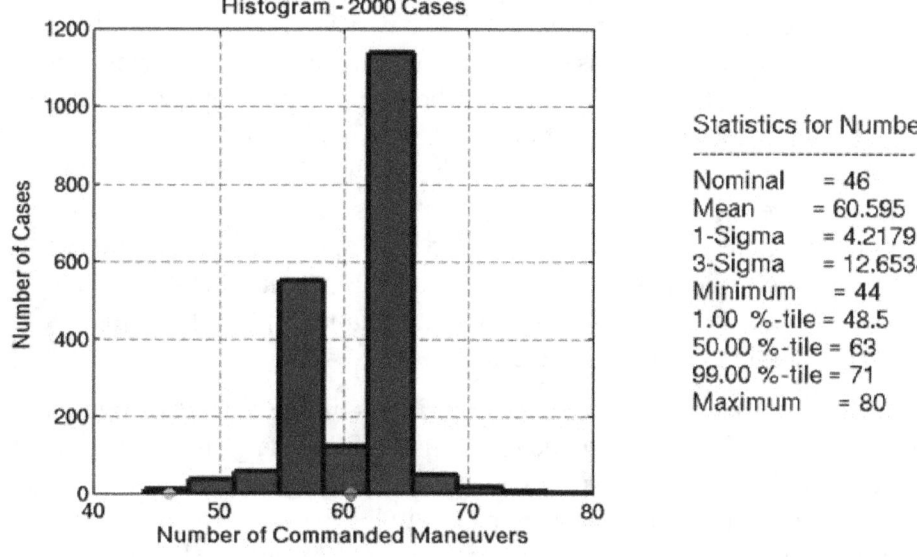

Figure 7.3-3. AADS Monte Carlo Results: Number of Maneuvers

Despite the range in number of maneuvers required per AADS mission run-out case, the actual variation in total ΔV was just under 3 m/s (3-sigma) across all 2000 cases, with a maximum case approximately 8 m/s higher than the nominal AADS case. This variation in total ΔV is small enough to be accommodated in a typical aerobraking ΔV budget, see Figure 7.3-4.

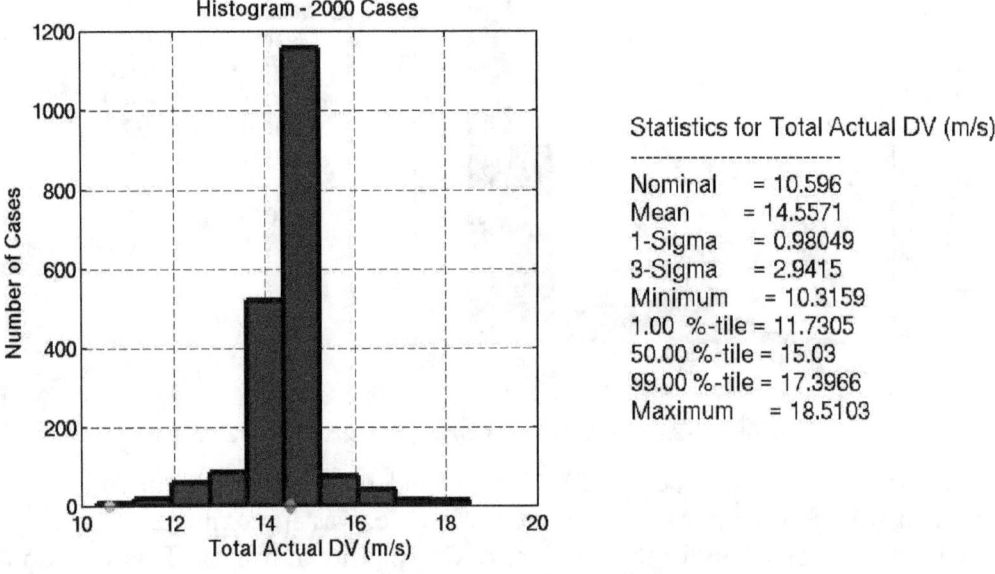

Figure 7.3-4. AADS Monte Carlo Results: Total ABM ΔV

Figure 7.3-5 shows an example of the total heat rate dispersion for a 2000 case Monte Carlo analysis utilizing the MarsGRAM atmosphere. The black dots show a comparison to the nominal mission, and the black dashed lines represent the upper and lower corridor limits used by the AADS. The typical immediate action line threshold of 100 percent above the upper corridor in this case would be located at a heat rate of 0.34 W/cm^2; thus all of the perturbed cases have heat rates that stay below the immediate action line. Figure 7.3-6 shows a histogram of the same data where the instances of heat rate in 0.01 W/cm^2 bins can be seen. The statistical mean heat rate (0.139 W/cm^2) is right at the middle of the corridor, and the 99 percentile heat rate is 19 percent above the upper corridor limit. These two plots, along with the other Monte Carlo results, suggest that AADS can accommodate significant variation in flight parameters while successfully completing missions with flight-like criteria.

Figure 7.3-7 shows the corresponding Monte Carlos periapsis timing errors through the mission. Increased timing error demonstrates propagated errors in the SP. The goal with the AADS weekly updates is to minimize the periapsis timing estimate. It is noted that even though the timing errors for the 1 and 99 percentile cases are nearly twice the periapsis timing errors in the nominal case, the performance capability of AADS is not compromised. This is evidenced by the previous discussion of the Monte Carlo analysis results for total ΔV, aerobraking duration, glide slope, number of maneuvers, and number of occurrences above the upper corridor limit.

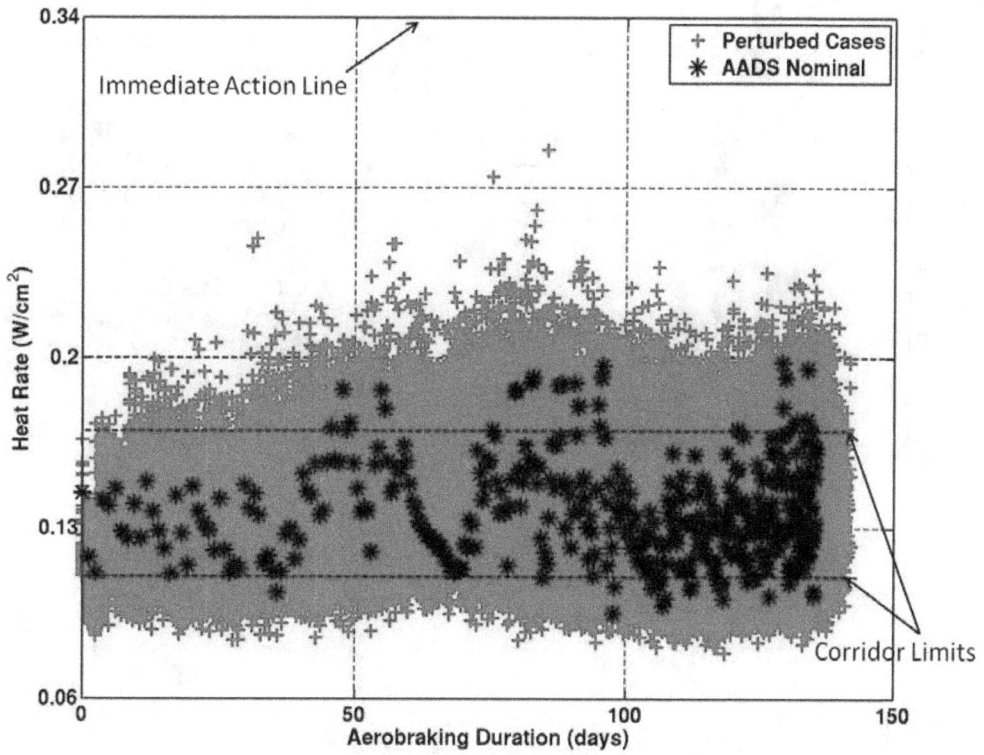

Figure 7.3-5. AADS Monte Carlo Heat Rate Distribution

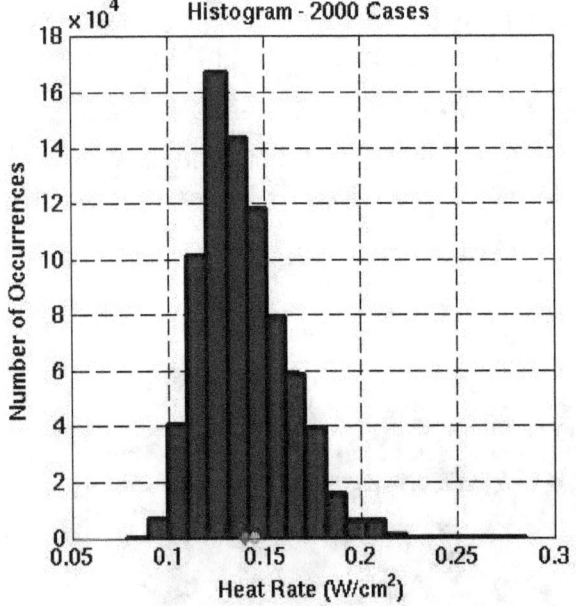

Figure 7.3-6. AADS Monte Carlo Results: Heat Rate Statistics

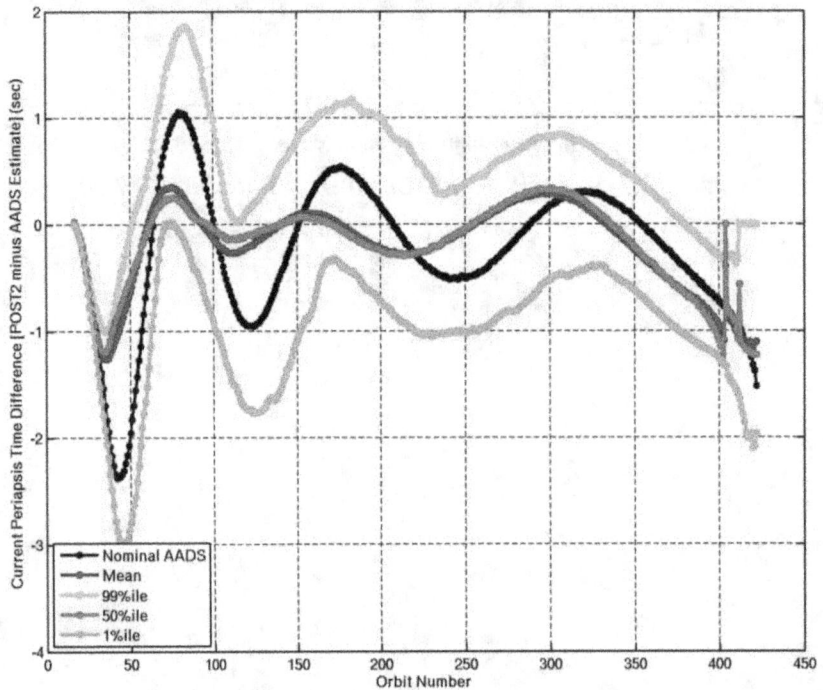

Figure 7.3-7. AADS Monte Carlo Results: Periapsis Timing Errors

Additional values that could be perturbed in future Monte Carlo analysis include, spacecraft mass, Mars dust storm variables and initial state conditions. Likewise, trade studies, using randomly selected normalized MGS, ODY, or MRO atmospheres, as described in Section 7.1.6 might provide insight to the performance using realistic profiles instead of MarsGRAM. These studies will determine which aspects of the AADS are most vulnerable.

7.4 Processor Load Analysis/Benchmarking

Demonstrating that processor capabilities can support the software requirements is critical to ensuring the AADS utility in a flight application. Although the code as written is not yet suitable for implementation on a flight processor, the software is mature enough to perform a preliminary assessment of the feasibility of its use in flight. It is time-consuming and expensive to port AADS to a real flight processor; this task is beyond the scope of Phase 2. However, simple benchmarking was done using the existing code running in a workstation that provides a rough order-of-magnitude assessment of the quantities of interest. Phase 1 identified characterizing the memory requirements and central processor unit (CPU) loading for the AADS a top priority.

During Phase 1, there was a concern that the memory requirements for the AADS would make execution in a flight processor difficult. At that time, the AADS memory requirements were largely due to the substantial buffers required for the attitude and acceleration data throughout the drag pass for later batch integration by the EE. These buffers were on the order of 3 megabyte (MB), which would complicate the memory handling in a flight processor. In addition to the buffered data from the spacecraft to the EE, there were smaller buffers of data from the EE to the AE that were approximately 0.5 MB. Because of the architecture changes made when converting the EE to the SP in Phase 2, these memory issues were resolved. Both of the large buffers noted above have been eliminated, and the memory footprint of AADS has been reduced by over an order of magnitude.

Assessing the processor loading for the AADS is essential to demonstrating the computational feasibility of AA. Actually testing the AADS on a flight processor is prohibitively expensive, but by performing timing tests of AADS running in a workstation, an assessment was made about the expected processor loading running in an embedded environment. The approach taken for this study was simply to run the AADS in the AAHFS simulation and to time each portion of the AADS that would correspond to a flight software task. These times obtained by the workstation testing were scaled based on the ratio of the clock speed of the workstation-to-flight processors to estimate the CPU time each task would require on a flight processor. Comparing this scaled time with the expected CPU time available for each task provided an estimate of the processor loading for each task. The testing described in this section used the AADS running on a workstation PC with a clock speed of 2670-MHz. As an example, the MESSENGER mission flight processors had a clock speed of 25-MHz, so the ground test times were scaled up by 107 times to estimate the equivalent CPU time required for a notional flight processor. The flight processor clock speed used in this analysis is considered conservative, as MESSENGER processors were purchased nearly a decade ago, and current-day missions are routinely running faster than 25-MHz.

In its current form, AADS is not explicitly decomposed into separate tasks as would be necessary for flight software. As a result, the benchmarking must only consider those functions that are related to an expected software task. This means that the benchmarking must carefully collect timing information on software functions that are only a part of the expected three tasks that comprise AADS. The tasks are: a high-rate task (notionally 10-Hz) for acceleration/attitude data processing, integration, and buffering; a low-rate task (notionally 0.01-Hz) for evaluation of the gravity models, integration of the spacecraft orbit, and preparation of the least-squares estimation matrices; and a batch process (executed once per orbit) that performs the orbit prediction and the subsequent AE and ME tasks. Each task will have different processor priorities and temporal allocations for completion.

The highest-rate process found in the AADS is the input data processing and high-rate integration function in the SP. There are few calculations associated with this software task, as it merely performs a trapezoidal integration of the acceleration data and buffers the resultant ΔV for use in the low-rate integrator. Although this task runs at a high rate, the processing required is minimal (a few multiplications and sums) and was not characterized, as it is by far the least concerning of the three processing tasks. It is well known that substantially more complicated algorithms are routinely running in spacecraft flight software at markedly higher rates (greater than 100-Hz), so no effort was devoted to benchmarking this task.

The low-rate integration task is the most challenging from a processor loading point of view. The SP is architected such that the low-rate integration attempts to step across a 100-s time period. During the exo-atmospheric flight, these integrations can routinely be performed in one or two integration steps (each of which requires 16 evaluations of the gravity models). During a drag pass, the integration steps can be much shorter (a few seconds each) requiring many more integration steps and many more gravity model evaluations. The processor loading will increase during each drag pass as the dynamics necessitate smaller integration steps to maintain tolerance, and because the variability in the atmosphere can require some back stepping in the integrator (essentially wasted integration steps). Benchmarking this task was performed at Mars, with a 21-degree and order gravity model, which is currently assumed to be the highest-fidelity gravity model considered for AADS testing (higher ordered gravity models are available, but do not offer significant enough increase in accuracy over the 21-degree and order field for purposes of aerobraking analysis). The results of this benchmarking throughout an entire Mars aerobraking mission are shown in Figure 7.4-1. Each drag pass sees a spike in the execution time for the low-rate integrator due to the shortened time steps. As the mission progresses, these spikes become more frequent as the orbital period drops and become more prolonged as the drag pass duration increases, but the peak value is not markedly affected. In the worst case, a flight processor would require approximately 1.6 seconds to complete the low-rate integration. Note that the resolution of the workstation timing data is 1-ms, so the scaled time is quantized at 107 ms. Since this task is assumed to run at 0.01 Hz, this reflects a processor loading (for this task only) of less than 2 percent. This result suggests that the processor loading for the low-rate integration is not an issue. If higher degree and order gravity models are used, there could be some increase in the expected peak processor loading, so any mission using the AADS would have to perform a CPU loading versus integration accuracy trade study.

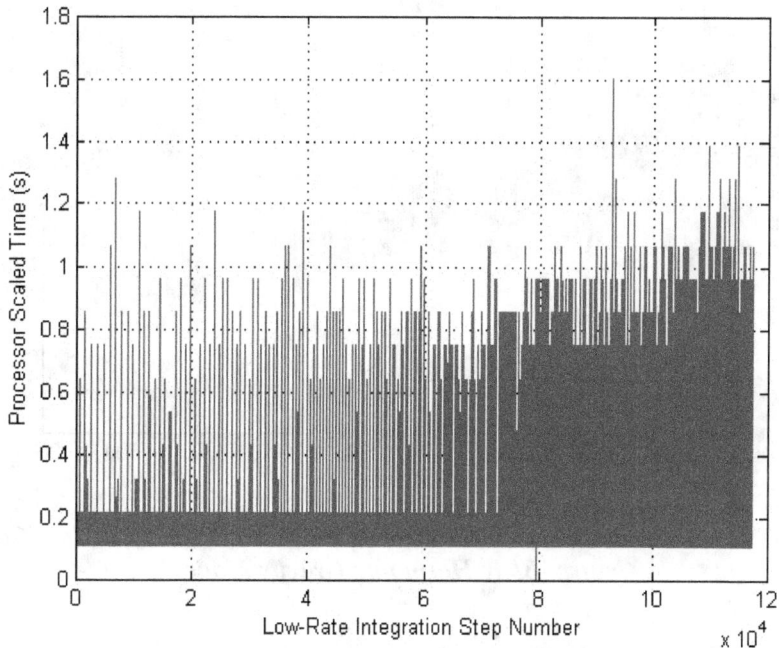

Figure 7.4-1. AADS Low-rate Integrator State Determination Scaled Time for Mars Case using 21x21 Central Body Harmonic Gravity Model

The AADS batch process that performs the orbital propagation and the atmosphere and maneuver calculations is by far the most computationally expensive task to complete and it will run at the lowest software priority. Despite the challenges that this task presents, the batch process is not expected to pose an issue for processor loading as there is a substantial amount of time to complete this task. It is assumed that the task will be triggered shortly after the drag pass exit, and must complete with sufficient time to configure the spacecraft for any corridor control maneuver. In the worst case, this affords the task approximately 25 percent of the minimum aerobraking orbit period to complete, or ~30 minutes. Timing tests were done to characterize the expected processor time required to complete this task, and the results are shown in Figure 7.4-2. For this task, the orbital propagation takes longer in the long period orbits, so the worst-case is early in the mission (when the task has much longer to complete). In all cases, the batch process would require much less than 1 percent of the available CPU, presenting no issue for the timely completion of this task.

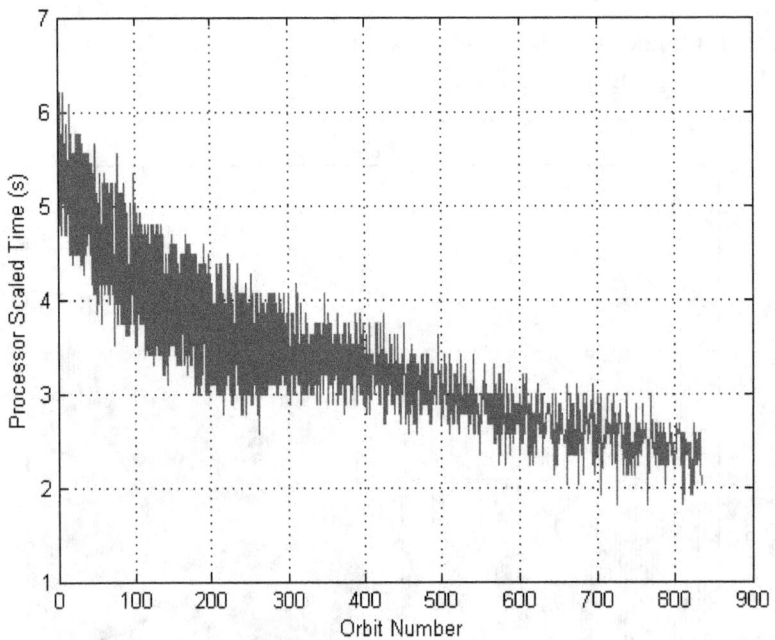

Figure 7.4-2. AADS Low-rate Integrator Orbit Propagator Scaled Time for Mars Case using 21x21 Central Body Harmonic Gravity Model

7.5 Other Trades and Sensitivity Analysis

7.5.1 Accelerometer Sensitivities

The analysis described in Section 7.2.2.1 is limited to cases without error sources that would affect the orbital integration. There are several notable error sources that could substantially affect the SP, including OD, accelerometer, and gravity model errors. Each of these error sources has been explored independently, but further analysis is needed to understand the coupling between these errors.

The SP relies on an accurate initial condition for its orbital integration, since there are no external measurements available to correct the integration. As a result, the integration is performed as a dead reckoning, and the only mechanism to correct drift in the onboard state is to perform a ground update. It is helpful to understand the uncertainty associated with the onboard state given an initial uncertainty associated with the ground update state. In other words, the navigation process that defines the ground update is uncertain, and this leads to greater uncertainty as the onboard state is propagated forward in time. The growth in this state covariance was computed using the Analytical Graphics, Inc. OD tool-kit and is shown in Figure 7.5-1 for a sample 7-day orbit propagation. The analysis assumes a typical initial uncertainty associated with the spacecraft state (1 m position uncertainty per axis and 1 mm/s velocity uncertainty per axis). The uncertainty growth shown in this figure demonstrates that after 7-day propagation, a typical initial state uncertainty can result in periapsis altitude errors of ~1 km. Cases were run with and without accelerometer errors. The two curves are largely indistinguishable, as the growth in the altitude uncertainty is dominated by the initial navigation error, and only mildly affected by the accelerometer error sources.

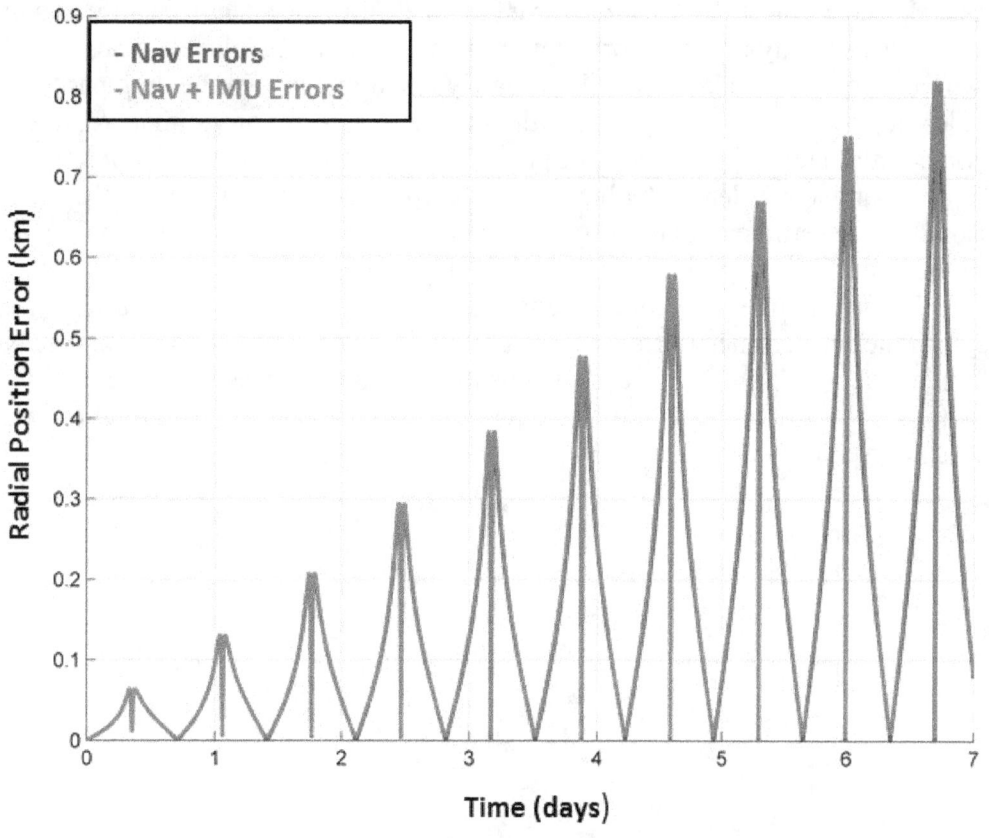

Figure 7.5-1. Ground Update State Uncertainty Propagated over 7-days

A limited assessment of accelerometer errors is described in this section, but a more extensive study is necessary to fully understand the impact these errors have on the state uncertainty, particularly for long propagation time spans. Because AAHFS has a full IMU model with adjustable noise, bias, alignment, and scale factor errors, this model is readily available for testing. Error parameters were set consistent with Northrop Grumman Scalable Space Inertial Reference Unit used for MESSENGER to quickly assess feasibility of AADS in the presence of realistic accelerometer error conditions. This is an identical analysis to that shown in the Phase 1 final report, although this testing used the updated Phase 2 AADS (running the SP instead of the EE). Qualitatively, the AADS is able to maintain reasonable corridor performance when using a 7-day ground update interval, as shown in Figure 7.5-1. Quantitatively, the drift in the onboard SP from the truth model (due to the accelerometer errors) is much more substantial than the error-free case, shown in Figures 7.5-2 and 7.5-3. These errors were not applied to previous study.

The drift in the orbit propagation due to accelerometer errors, presented in Figures 7.5-2 and 7.5-4, is a notional example. Most of the accumulated error in the SD process of the SP is due to uncompensated accelerometer bias, as this error causes drift in the state. Because this bias is fixed in the spacecraft body frame, its orientation with respect to the planet-relative velocity direction will also remain fixed. This will ultimately result in a bias in the drag pass ΔV estimation, leading to nearly monotonic growth in the periapsis timing estimate (and to a lesser degree, periapsis altitude). The envelope of possible orbit errors were not characterized by this analysis, as only sample accelerometer biases were used that characterize one possible outcome with a reasonable accelerometer bias. The possible biases errors themselves map into an ellipsoid in the body frame, which in turn translates into an error ellipsoid in the ΔV estimated at each drag pass. Further analysis would be required to characterize the worst-case state errors associated with the accelerometer errors. The AAHFS is sufficient to conduct this analysis, but it was not undertaken in Phase 2 due to time constraints. This analysis would require numerous simulations to perform the Monte Carlo or sigma-point analysis and each simulation requires more than 12 hours to complete.

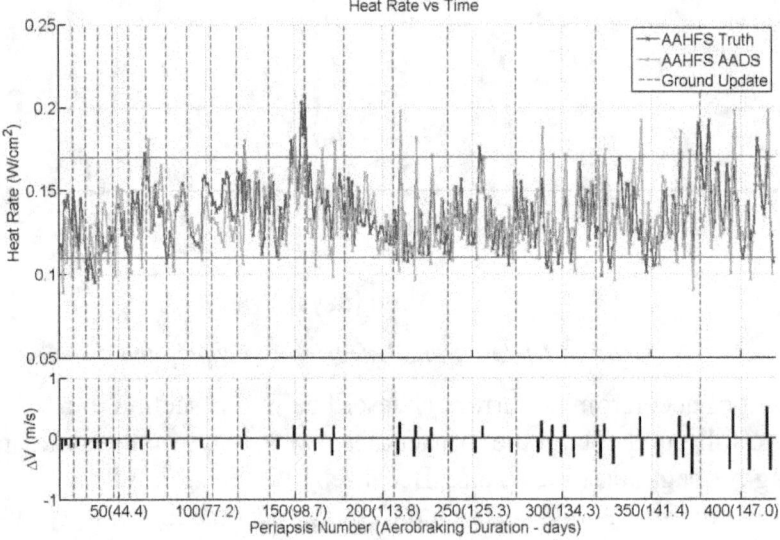

Figure 7.5-2. Corridor Performance for Phase 2 AADS with 7-day Ground Update and Accelerometer Errors

NESC Request No.: TI-09-00605

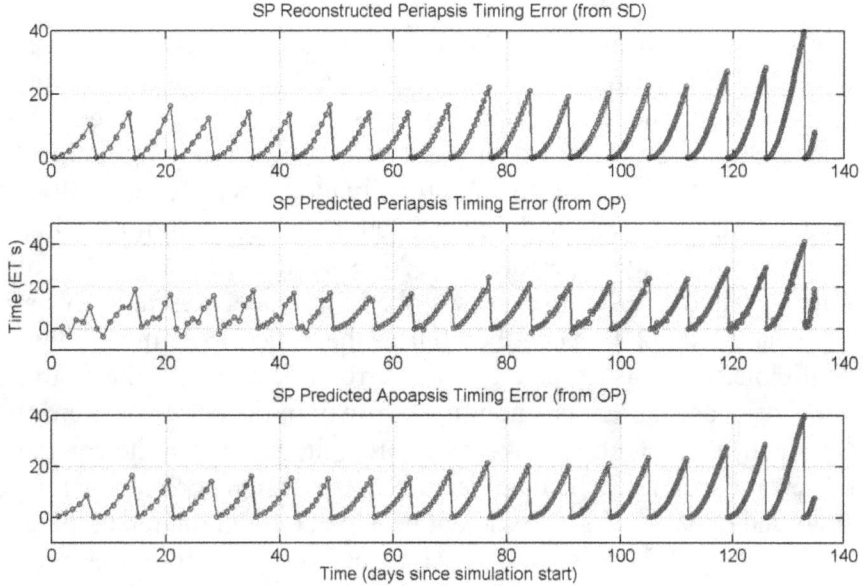

Figure 7.5-3. Apses Timing Performance for Phase 2 AADS with 7-day Ground Update and Accelerometer Errors

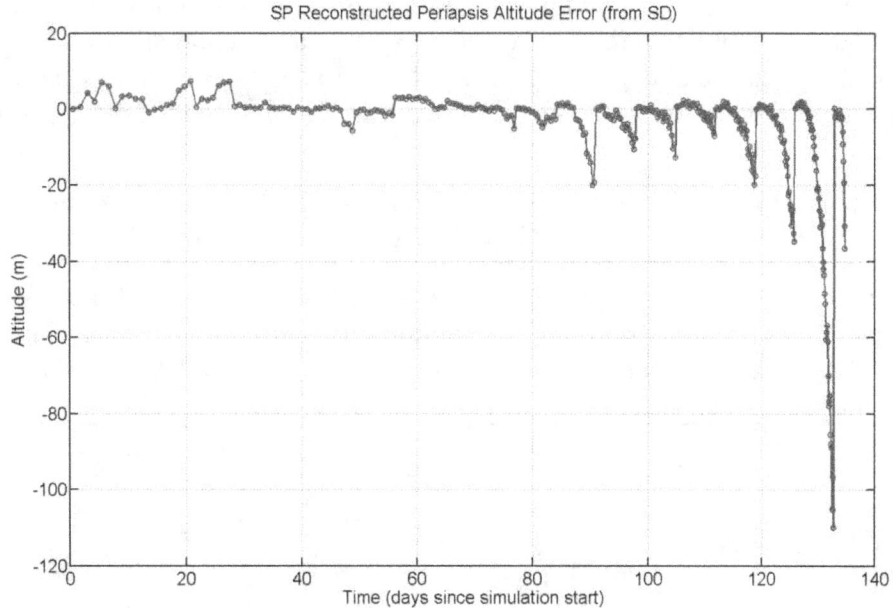

Figure 7.5-4. Periapsis Altitude Performance for Phase 2 AADS with 7-day Ground Update and Accelerometer Errors

7.5.2 Gravity Model Sensitivities

All simulations described above and in the Phase 1 report used identical gravity models in the truth model and the AADS. Whatever gravity model would be used in flight would not perfectly represent the true gravity conditions, and one way to assess the performance of AADS to gravity modeling errors is to introduce a difference between the AADS gravity model and the truth model. This study gives some insight into what fidelity model is required in flight to maintain good SP performance, in terms of maintaining an accurate periapsis timing estimate and altitude. Table 7.5-1 summarizes the analysis performed at Mars using the EML version of AADS, running with a 7-day ground update. In all cases, the truth model central body harmonic gravity model used a degree and order 21. When the SP uses the same 21x21 model, the errors after 7 days remain negligible, which is expected for this error-free case. As the degree and order of the gravity model is decreased, there is a nonlinear growth in the periapsis metrics. The lowest-fidelity gravity model tried (3x3) shows errors that are quite large, and the corridor control for that case showed several prolonged excursions outside the design corridor that were only corrected with a ground update. This assessment provided a rough understanding of the gravity model fidelity that is required for reasonable AADS performance. From these results, it appears that low-fidelity gravity models are insufficient to ensure adequate corridor performance and given the modest increase in SP execution times required for higher-fidelity model evaluation, a good balance can be achieved between processor loading and integration accuracy.

This analysis would need to be repeated for any host mission, as it depends heavily on the central body, the reference trajectory path relative to central body gravity anomalies, and the accuracy of the gravity model at a particular degree and order. These trades can be quickly performed using AAHFS or POST2, as the software parameters that define the gravity model fidelity are easily adjusted.

Table 7.5-1. Impact of Gravity Model Errors on SP Accuracy and Timing for 7-day Ground Update at Mars when Compared Against Fixed 21 degree and order Gravity Truth Model

AADS Mars Harmonic Gravity Model Degree and Order	Worst-case Reconstructed Periapsis Timing Error (s)	Worst-case Periapsis Altitude Error (m)	AADS Execution Time on Ground Workstation (ms)
21x21	0.15	3	15
17x17	2	120	10
13x13	6	320	6
9x9	9	680	5
3x3	100	2400	3

7.6 AADS Performance Assessment

During Phase 1, extensive testing of AADS was performed using an engineering model of the atmosphere with dispersions to demonstrate feasibility of the algorithm. Phase 1 concluded that AADS had passed the initial feasibility tests, but needed to be rigorously tested in an operational flight-like environment. In Phase 2, AADS performance was verified using a more stressing aerobraking scenario than was experience by the Mars ODY aerobraking mission which was the most aggressive (maintaining the least amount of thermal margin) to date. To test the AADS in the most realistic sense, an ORT was used. An ORT is designed to evaluate the ability of a team to perform all mission critical tasks prior to the commencement of a mission. Because there is no identified flight vehicle for this testing, the Phase 2 demonstration is not an ORT in the most proper sense. However, the concept of the ORT provides an ideal testing environment for the AADS. The key to a successful ORT is the simulation and processes that recreate the flight and operational experience. By focusing the AADS Phase 2 analysis as an ORT at Mars, the simulation could take advantage of observed density profiles from the three previous aerobraking missions at the planet (MGS, ODY and MRO) and demonstrate the AADS capabilities in a realistic mission operations setting. The performance and advantages of the AADS are evaluated by running two aerobraking ORTs simultaneously: one in the historical ground-based manner (ORT1) and one using AADS (ORT2). So, not only is the AADS prediction and reaction being compared with what happens on a spacecraft (ORT2), but it is also compared to the decisions that a ground-based team would have made with a similar spacecraft and environment (ORT2 versus ORT1). Additional variations of ORT2 were run after the initial assessment to evaluate additional AADS features including atmosphere dispersions in the maneuver logic (ORT2a) and the effect of using constant corridors (ORT2b).

The ORT scenario was deliberately challenging, in that it was designed to aerobrake from an 18-hour period orbit to an orbit with a 2800 km apoapsis altitude in ~70 days with a margin of only 100 percent (see the definition provided in Section 7.2.1), the same as for Mars ODY. The ground-based team met each simulation "day" during ORT1 (e.g., after each 24 hours of simulated mission elapsed time), and decided on the maneuver strategy for that day. Every week of mission elapsed time, the ground-based team also met, examined the latest mission run-out, and decided if the corridor needed to be adjusted. The AADS Team met only weekly. Both ORTs successfully achieved the final mission requirements, but the AADS Team required fewer resources, namely no daily meetings, and only needed weekly updates from the Atmosphere Team for the mission run-outs.

The reference mission, operational procedures, simulations, and results for the Phase 2 aerobraking ORTs are described in Section 7.6.1.

In addition to the ORT, AADS was used for additional analyses, including simulated replication of the ODY and MRO aerobraking missions. Discussions of these simulations are provided in

Section 7.6.2 and 7.6.3, respectively. In summary, the AADS has shown to be a robust technique that offers more flexibility than the current ground-intensive method.

7.6.1 Mars ORT

As previously mentioned, an ORT is designed to evaluate the ability of a team to perform all mission critical tasks prior to the commencement of a mission. The key to a successful ORT is the simulation and processes that recreate the flight and operational experience. Historically, real-time aerobraking operations involve dozens of individuals working at several NASA Centers, universities, and industry locations to download data, analyze the environment and spacecraft, to make maneuver decisions and finally, upload data to the spacecraft.
Table 7.6-1 shows a listing of the typical aerobraking operations teams and their tasks during ground-based aerobraking operations. A successful aerobraking ORT must realistically "simulate" a core of each of these elements.

Table 7.6-1. Aerobraking Operations Ground-based Analysis Teams

Team	Purpose
Spacecraft	Evaluate and report on spacecraft health and safety
Navigation	Build and upload maneuver decision to spacecraft
Flight Mechanics	Perform analysis to inform maneuver decisions and evaluate mission performance margins
Atmosphere Advisory	Evaluate atmosphere conditions and report effect on mission and maneuver decisions
Thermal	Monitor spacecraft temperatures and infer heating for each orbit
Accelerometer	Convert accelerometer data into atmosphere data
Aerodynamics	Compare flight instrument data with model data, make updates as needed

For the Phase 2 evaluation of the AADS, ORTs 1 and 2 utilized the Accelerometer, Flight Mechanics and Atmosphere Advisory Teams. The spacecraft is assumed to be healthy and safe unless otherwise indicated by derived aeropass maximum heat rate values. The navigation and accelerometer team tasks will be included in the spacecraft flight simulation (SFS). It is assumed that the SFS or "truth" simulation accurately represents the thermal and aerodynamic properties thus eliminating the need for representation from those teams during the ORTs.

7.6.1.1 ORT Reference Mission

Both ORTs attempted to follow the same ODY-like reference mission. The reference mission, initialized on April 9, 2009, was designed to reduce an initial orbit of 18 hours (270,000 km apoapsis altitude) to 3 hours (2800 km apoapsis altitude) in approximately 70 days. The reference mission ends at 2800 km because the AADS is not currently designed to accommodate collision and avoidance (COLA) considerations or the walk-out scenario, nor is it equipped with a lifetime constraint algorithm. Therefore, neither ORT will consider this late phase of aerobraking. The mission assumes a walk-in from 200 km to approximately 95 km periapsis altitude over six orbits, performing three maneuvers (one every other orbit) until the maximum heat rate is near the target value of 0.112 W/cm^2. The reference design mission simulation then calculates the precise ΔV required to obtain the target heat rate on each subsequent pass through the reference atmosphere (MarsGRAM2010). A plot of the periapsis heat rate versus time for the reference mission is shown in red in Figure 7.2-1 in Section 7.2.1. The simulation provided a reference glide slope, or LTST of the ascending node versus time (as aerobraking duration in days or orbit number) plot, that was be used by the ground-based and AADS simulated missions to evaluate mission progress. The goal of both missions was to end at the reference mission final LTST of 10:46 hours +/-15 minutes. The 15-minute margin was a requirement set by the science instrument operability for ODY, but it also provided flexibility in terminating aerobraking during the end-game phase of the mission. The reference glide slope, with the 15-minute target bounds, is shown in red in Figure 7.6-5 in Section 7.6.1.4.

All simulations used a spacecraft model with the same reference area and aerodynamics as the MRO spacecraft, but the mass of the vehicle was reduced from ~1300 kg to 850 kg in order to lower the ballistic coefficient and allow the reference mission to complete the main aerobraking phase in the desired 70 days.

7.6.1.2 Atmosphere Team Analysis

The Atmosphere Team function, preserved from past aerobraking missions, was utilized in both ORTs. Since neither ORT simulated *a priori* knowledge of the atmospheric density before entry, the MarsGRAM 2010 model was used for predicting the environment. The model assumed smooth bell-shaped density curves that increase as altitude decreases. However, the actual flight atmosphere in the SFS or "truth" simulation, taken from the ODY mission, resulted in a difference between actual and predicted flight density profile. An example of the difference between the model and observed density profiles is shown in Figure 7.6-1. The Atmosphere Team monitored these differences and provided model updates (different for each ORT) in the form of a density multiplier (MarsGRAM parameter A0) to be applied only to those analyses predicting future aerobraking performance. The multiplier was derived from taking the ratio of the integrated density from the SFS (e.g., blue line in Figure 7.6-1) simulation to the integrated density from the MarsGRAM prediction (using A0=1). The multiplier was evaluated after each orbit, but was only changed daily for ORT1 and weekly for ORT2. Figure 7.6-2 shows the multiplier for each orbit and the value used in the simulation for both ORTs.

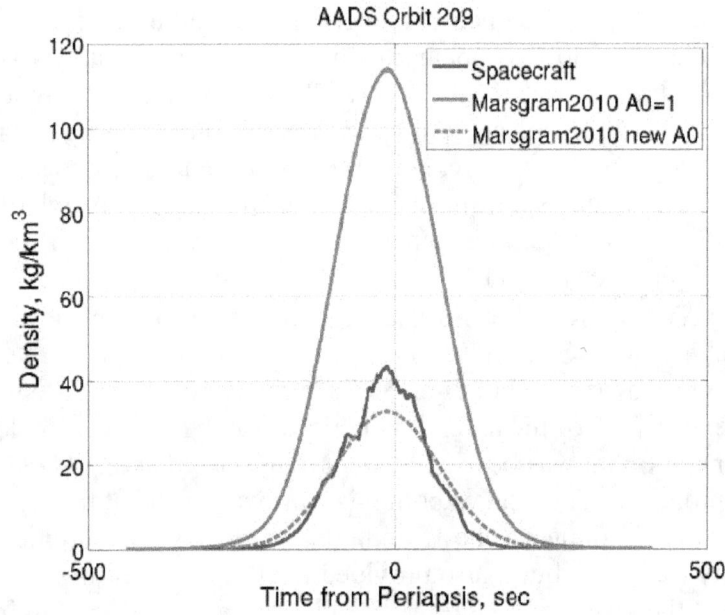

Figure 7.6-1. Profiles of Observed Density (blue), Model Density with A0 =1 (solid red) and the Model Density Scaled to Match area under Observed Density Profile (dashed red) for a sample AADS Mission Orbit 209

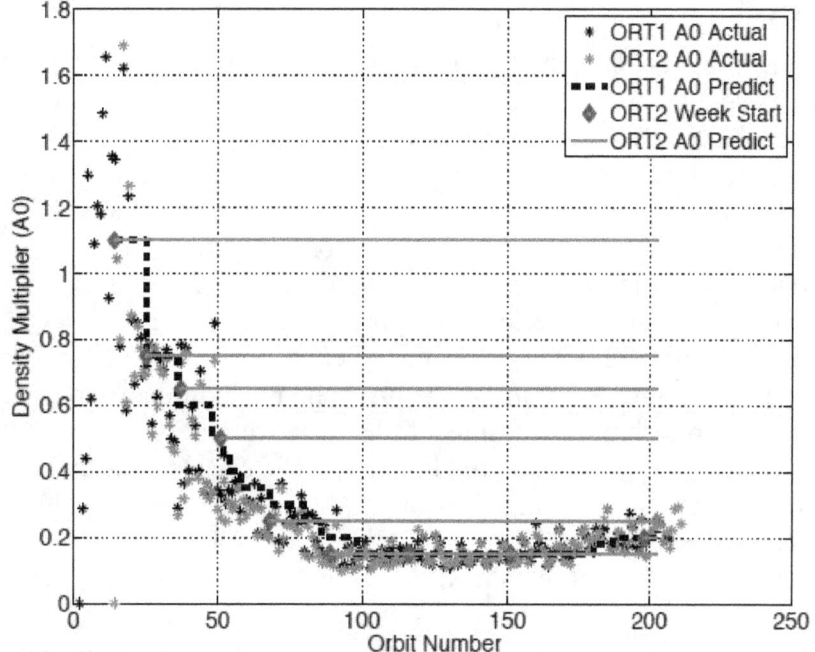

Figure 7.6-2. A0 Multipliers Observed and used in the Simulations

7.6.1.3 ORT1: Historical Ground-Based

To offer a realistic comparison to the operational advantages of using the AADS, a historical ground-based ORT was used to baseline the "typical" amount of work (number of tasks and time needed to perform them) required during aerobraking for evaluating spacecraft data and recommending maneuvers (e.g., to perform the tasks that AADS performs automatically). The walk-in phase of the mission, which is similar in both ORTs, executed three maneuvers over six orbits to gradually dip into the atmosphere, increasing the peak heat rate to be within the corridor. The corridor was determined in the operational mission run-out, described in Section 7.2.1, which resulted in an initial guess for the corridor to be 0.085 W/cm^2 for the lower bound and 0.14 W/cm^2 for the upper bound. The goal of the ground-based Flight Mechanics Team was to utilize the atmosphere predictions from the Atmosphere Team and perform analysis to determine when and what size of maneuvers should be made to keep the spacecraft heating within the required corridor in order to ensure that the mission followed the predetermined reference glide slope.

During ORT1, the Flight Mechanics and Atmosphere teams performed two types of analyses to support maneuver decisions: daily and weekly. A description of both analyses is provided below.

ORT1: Daily Analysis

Daily analysis evaluates the potential effect of all maneuver menu options and utilizes the ABM sweep simulation described in Section 7.2.1. The simulation implements the density multiplier (A0) from the Atmosphere Team and the results were used to make a maneuver recommendation to the mission manager, who ultimately decides if a maneuver will be performed (i.e., uploaded to the spacecraft). During a real mission, the daily analysis begins by evaluating the "no maneuver" option, which is then used to determine a subset of maneuver options to evaluate. The likely optimal maneuver option is then further evaluated using Monte Carlo analysis to assess how the 1 and 99 percentile trajectories compare to the corridor. However, in the interest of time for the condensed Phase 2 ORT1 schedule (where a day of aerobraking was simulated in 1 hour), all maneuver options were run simultaneously and no additional Monte Carlo analysis was performed. Figure 7.6-3 shows the results of the maneuver options on orbit 95 starting from the state at apoapsis of orbit 93. The thick red line at 0.28 W/cm^2 denotes the immediate action line and the constant red lines at 0.085 and 0.14 W/cm^2 denote the upper and lower corridor targets, respectively. The vertical black-dashed lines denote the planning period (the next 48 hours) for which the spacecraft must remain within the corridor limits in the event that no maneuver can be uploaded. The potential effects of performing each maneuver (based on the MarsGRAM atmosphere and the density multiplier provided from the Atmosphere Team) are shown. As expected, a large negative maneuver, which would decrease the periapsis altitude (increase heat rate indicator), would likely result in heat rates above the upper corridor. Likewise, a large positive maneuver (i.e., 0.6 m/s) would likely raise periapsis altitude such that the resulting heat rate on subsequent orbits would be below the lower corridor. In this particular example, performing no maneuver would result in the spacecraft heat rates falling below the lower corridor throughout the planning period and is highlighted using a black-dashed line. For this case, the ground team chose a maneuver of -0.10 m/s during ORT1.

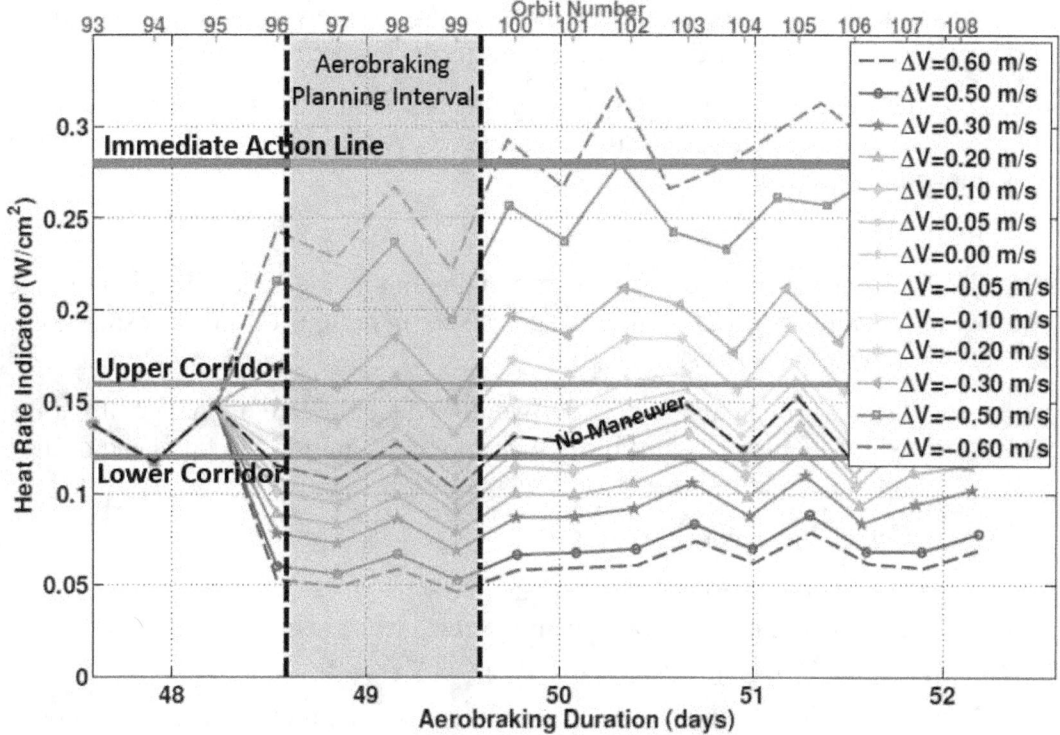

Figure 7.6-3. ORT1 Daily Analysis

ORT1: Weekly Analysis

Like the daily analysis, the ground-based weekly analysis was based on past aerobraking mission flight experience. The purpose of the weekly analysis was to evaluate the margin to the targeted final orbit defined by the reference mission. This margin is determined by flying three simulated missions that target exactly the heat rate of the upper, middle, and lower corridor. The simulations are initiated with the latest spacecraft state at apoapsis and end at 2800 km apoapsis altitude (the point at which the use of AADS will cease). The results are then compared to the reference mission to determine if the corridor needs to be adjusted in order to meet the desired end conditions. Figure 7.6-4 shows the orbital period results of the three simulations that are initialized with an apoapsis state on day 16. Flying a mission near the upper corridor will result in a mission approximately 10 days shorter than the reference. However, flying the rest of the mission near the lower corridor will result in the aerobraking mission taking about 10 days longer than the reference. Figure 7.6-5 also shows the LTST margin for this same simulation, where the horizontal red lines denote the desired target range at the end of the mission (EOM). In this case, if the mission continues to fly near the lower corridor, the mission will miss the target LTST of the ascending node by 20 minutes. Therefore, a near-term maneuver strategy should focus on remaining in the upper part of the corridor for a few orbits in order to pull the lower corridor mission back into the final mission limits during the next weekly run-out.

The team may also choose to change the corridor limits based on the results of the weekly analysis. Depending on the results of the run-out, several options exist for modifications to the maneuver strategy for the upcoming week, such as:

1. Modifying the upper corridor.
    a. Increasing it reduces the margin to the immediate action line, but may serve as a viable option to fly more aggressively in regions of the atmosphere that are known to have less variability.
    b. Decreasing the upper corridor would reduce the rate of change of the LTST, which would be useful if the spacecraft was getting too far ahead of the reference glide slope.
2. Modifying the lower corridor.
    a. Increasing the lower corridor can allow the mission to move closer to the reference if the mission is behind schedule. It also can result in additional maneuvers being required.
    b. Decreasing the lower corridor can lengthen the mission duration and reduce the number of required maneuvers.

For ORT1, the corridor was modified three times throughout the 70-day mission. Figure 7.6-6 shows the corridor with both the predicted and actual heat rates observed on each orbit. Because of the lower than expected densities (A0 were near 0.2 for most of the mission), much of the mission flew near or below the lower corridor (see the blue dots in Figure 7.6-6). To remain on the target glide slope, the lower corridor was raised. Additionally, because of the low densities, the ground team also decided to raise the upper corridor. This effectively reduced the thermal margin, but did so to make effective use of the atmosphere when risk was perceived to be low.

Near the EOM, when the orbit period is less than 3 hours, there are 8 orbits occurring between maneuver opportunities. Maneuvers are determined based on the latest atmosphere trends provided from the Atmosphere Team and the ABM sweep. However, in the ground-based ORT1, the variability in the atmosphere near the EOM was not captured in the model and therefore the heat rate on three orbits violated the immediate action line before a maneuver could be performed to raise the periapsis altitude effectively reducing the periapsis heat rate.

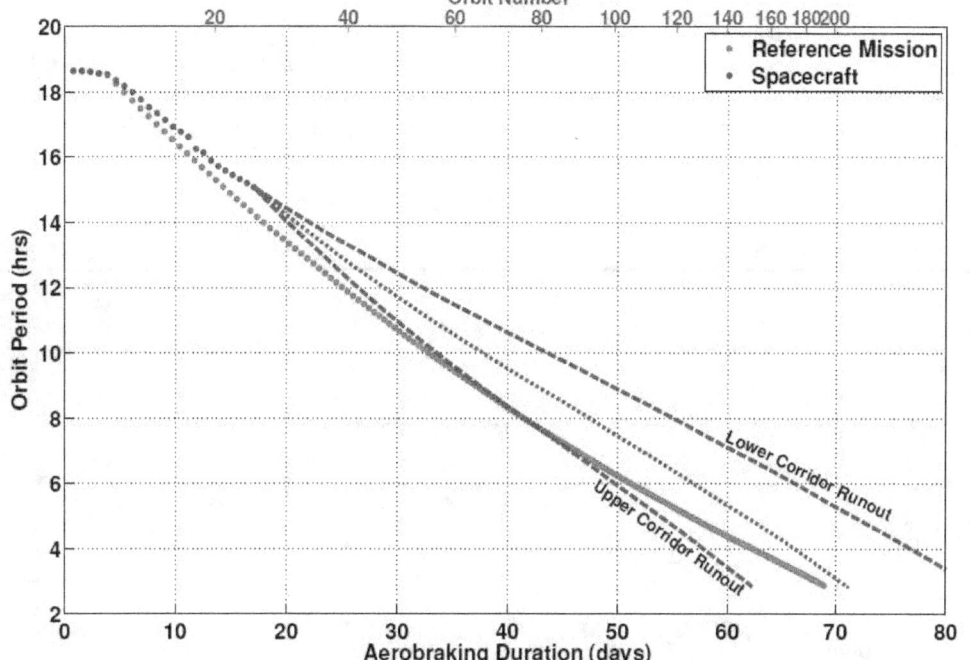

Figure 7.6-4. ORT1 Weekly Analysis: Orbit Period

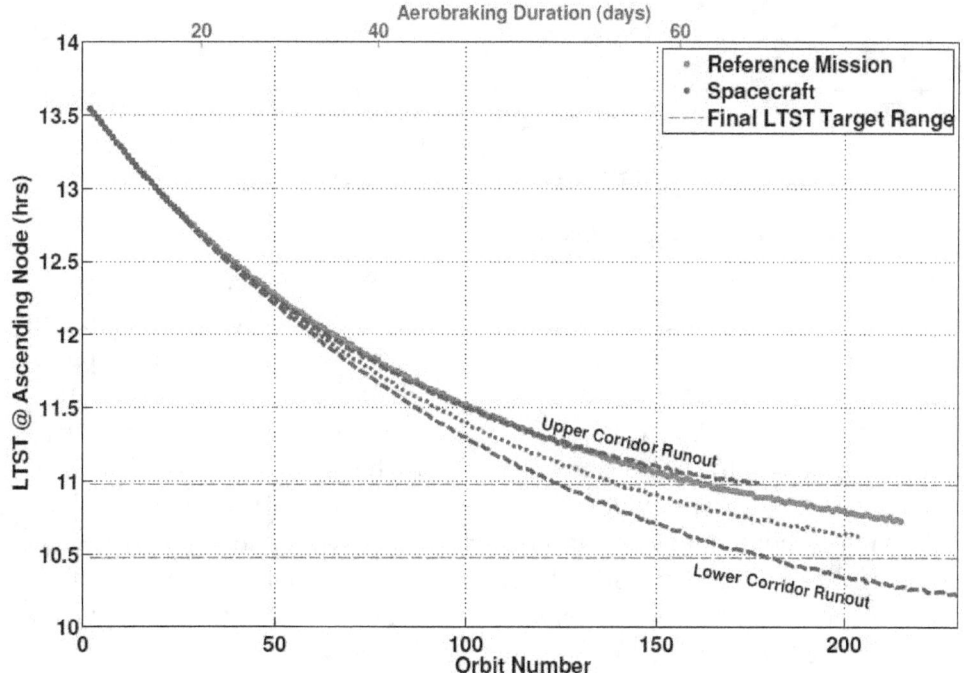

Figure 7.6-5. ORT1 Weekly Analysis: LTST

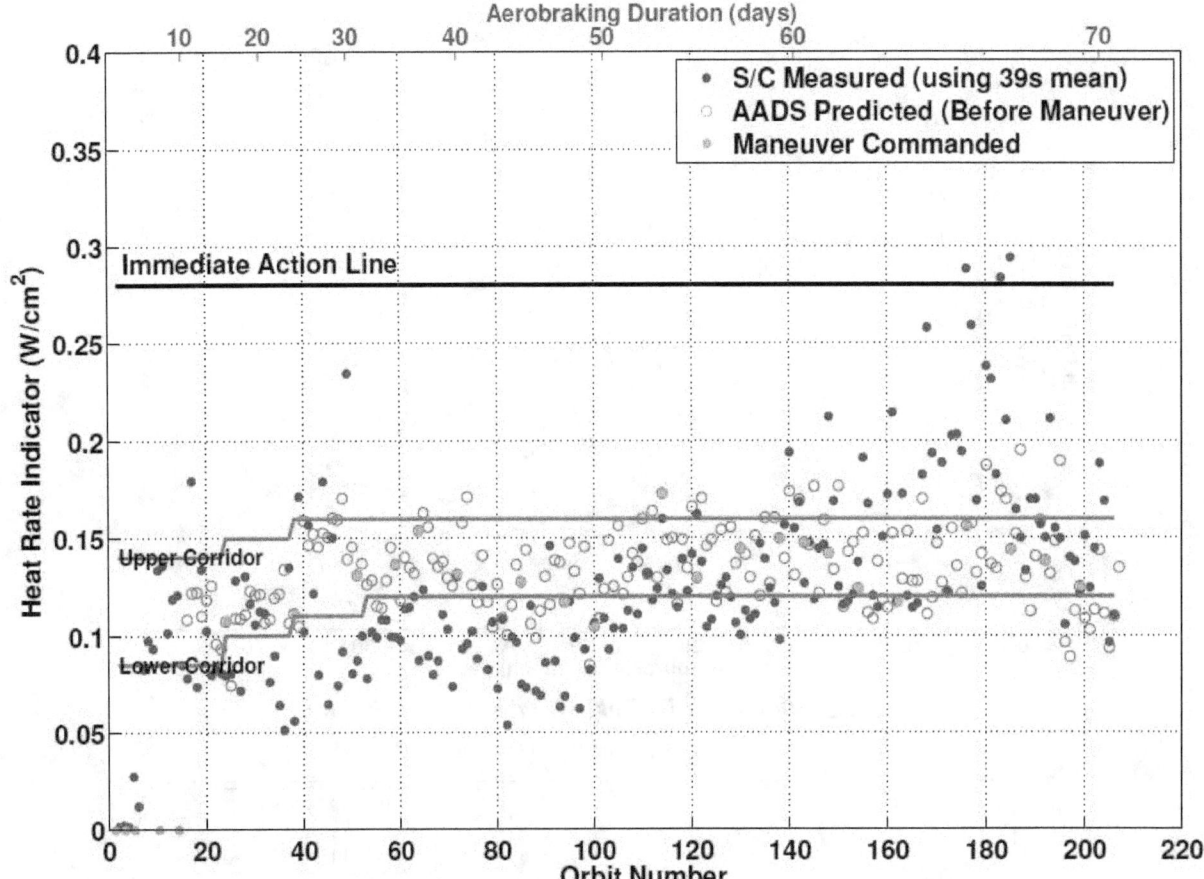

Figure 7.6-6. ORT1 Heat Rate versus Days of Aerobraking

ORT1 Maneuver Decision-making Process

During the ODY aerobraking mission, a maneuver could be performed only once per day. However, if a maneuver was made, it was designed such that under nominal conditions the next maneuver could not occur within the next 48 hours unless a maneuver was required the next day to protect the spacecraft. This policy mitigated the work of the ground team to build, test and upload maneuver commands. A similar policy was adopted for ORT1. The SFS was stopped every 24 hours. The state at the nearest apoapsis was written to file and provided to the ORT1 Flight Mechanics Team as the spacecraft state file. The team would select on orbit apoapsis within the following 24 hours that was eligible to perform a maneuver should it be deemed necessary. In keeping with ODY and MRO fight experience, whenever possible, the orbit was selected to occur during the prime shift (between 8 a.m. and 5 p.m. local time). Once an apoapsis was designated as available for a maneuver, the daily analysis was initiated. The Flight Mechanics Team would consider the effect of each maneuver menu option on the propagated trajectory over the next 48 hours. The maneuver decision was often straightforward, as in the case when the no maneuver option kept the spacecraft within the corridor. Other times, the no maneuver case would fall outside the corridor on one or more orbits, at which time the team had to evaluate the effectiveness of the overall aerobraking strategy, including:

1. How well the mission was following the reference glide slope; if the mission is running behind, the strategy may be to perform maneuvers to keep the maximum heat rates near the top of the corridor.

2. How the orbital mechanics of the trajectory impact periapsis altitude on subsequent orbits; as the periapsis latitudes dropped, periapsis altitudes increase (due to the Martian gravity field for the latitudes flown in the ORT) resulting in the desire to remain at the lower part of the corridor because natural precession would increase heating on subsequent orbits.

3. Number of corridor violations; if only one orbit exceeded a corridor did it warrant using a smaller/larger maneuver?

4. Looking ahead; what would the situation look like tomorrow? Should a small maneuver today be traded for a larger maneuver tomorrow (or vice versa)?

All of these criteria were used to determine whether or not a maneuver would be made. In the end, ORT1 opted to perform 22 maneuvers requiring 4.8 m/s ΔV. The solid green circles in Figure 7.6-6 denote the orbits where maneuvers were commanded. The total ORT1 mission duration was 70.4 days. The results of the decisions made for ORT1 are compared to ORT2 in Section 7.6.1.5.

7.6.1.4 ORT2: AADS with the SP

Because AA is only intended for use in the main phase of aerobraking, and because the AE requires a data archive used to make atmosphere predictions, the ORT2 AADS simulation begins at orbit 14 of the reference mission, approximately 10 days after the first walk-in orbit.

The AADS eliminates the need for nearly all of the teams listed in Table 7.6-1 because the software moves the maneuver decision-making process onboard the spacecraft. However, while AADS eliminates the need for all of the daily analysis, to evaluate mission margin, the Flight Mechanics Team still performs the weekly analysis on the ground to determine whether corridor adjustments are required to meet the overall aerobraking constraints. The weekly analysis occurs in the same manner as described for ORT1. The Atmosphere Team provides an update to the model A0 based on the comparison of the observed and modeled densities from the previous week and considerations for the week ahead. The updated A0, together with the spacecraft apoapsis state at the end of the week, is used to initiate three simulations — one each targeting the heat rate of the upper, middle, and lower corridor. The results are used to determine what changes, if any, should be made to the corridor over the next week to keep the mission on track (i.e., with respect to the glide slope).

Due to the similarity with ORT1, week 3 run-out analysis, shown in Figures 7.6-4 and 7.6-5, plots are not provided for ORT2 at the same point in the mission. As with ORT1, ORT2 analysis at the end of week 3 indicated that the lower corridor could be raised slightly to speed up the mission. Eventually, starting after week 4, the upper corridor was also lowered, effectively increasing the margin to the immediate action line, a move contrary to the decisions made in ORT1.

Figure 7.6-7 shows the heat rate corridor plot for the ORT2 simulated mission. It is noted that the heat rate on orbit 39 and 48 exceeded the immediate action limit line. In ORT2, maneuvers were commanded based only on the predicted value of the heat rate at the next periapsis (open red circles in the plot). If the heat rate was within the corridor, no maneuver was performed, and if outside the corridor a maneuver was performed to place the next periapsis heat rate at the specified point within the corridor. As a note, ODY also had one orbit that reached the immediate action line. The maneuver logic and pop-up capability described in Section 7.1.5 are fully operational. However, the maneuver logic based on the 3-sigma high-density prediction was not active in the simulation, but the 3-sigma heat rates were calculated and are shown in Figure 7.6-8.

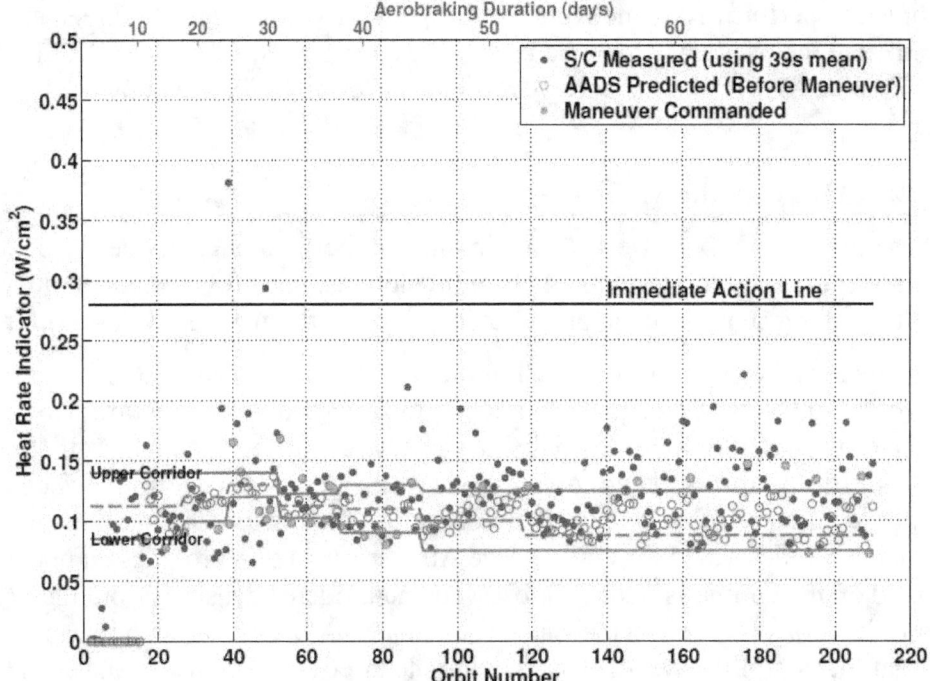

Figure 7.6-7. ORT2 Heat Rate Corridor

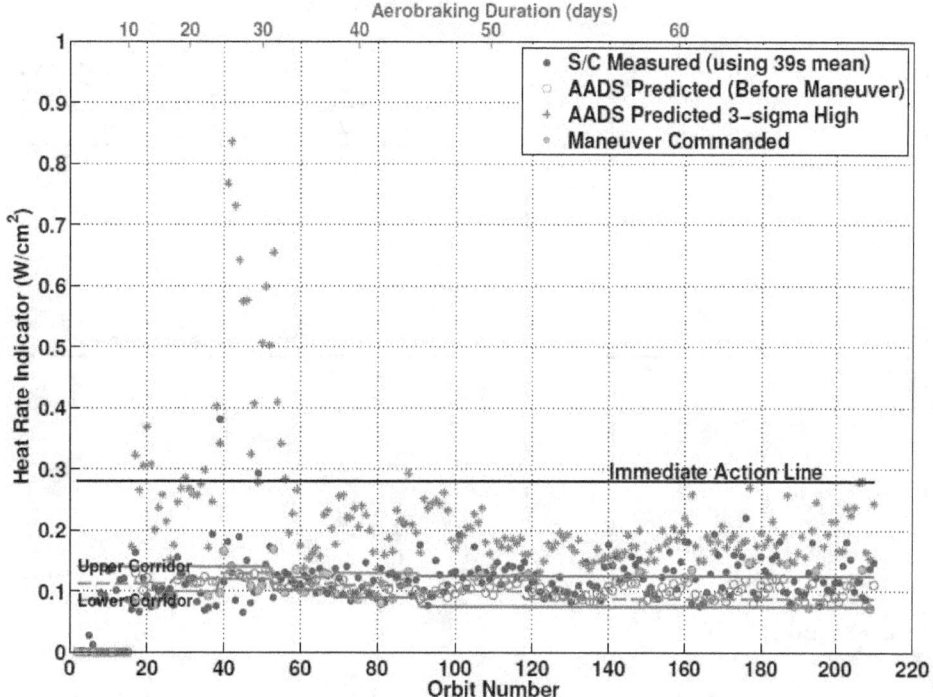

Figure 7.6.8. ORT2 with the 3-Sigma High Heat Rate Predictions

For a region of the atmosphere with considerable uncertainty, between orbits 40 and 60, the 3-sigma high-predicted heat rates are well above the immediate action line. Therefore, ORT2 was re-run using the maneuver logic, which considered the 3-sigma values (ORT2a). The mission was completed utilizing a similar corridor profile, while successfully protecting against the immediate action limit violations. The corridor selections in ORT2a are similar to those used in the original ORT2, but the overall mission profile is different because of the additional biasing maneuvers. A plot of the heat rates observed for ORT2a are shown with the corridors and predicted heat rates in Figure 7.6-9. The figure also denotes the orbits for which the AADS commanded a maneuver. Also, it should be noted that for this analysis (and all subsequent analyses), the pop-up logic was enabled, but no pop-up maneuvers were ever determined to be necessary.

It is important to note that while ORT1 experienced periapsis heat rates that exceeded the immediate action limit near the EOM due to the ability to only make a maneuver once per day, the AADS ORTs did not because a maneuver can be made each orbit. AADS can maintain periapsis heat rates closer to the corridor limits in similar atmosphere conditions. This is another way AADS reduces mission risk.

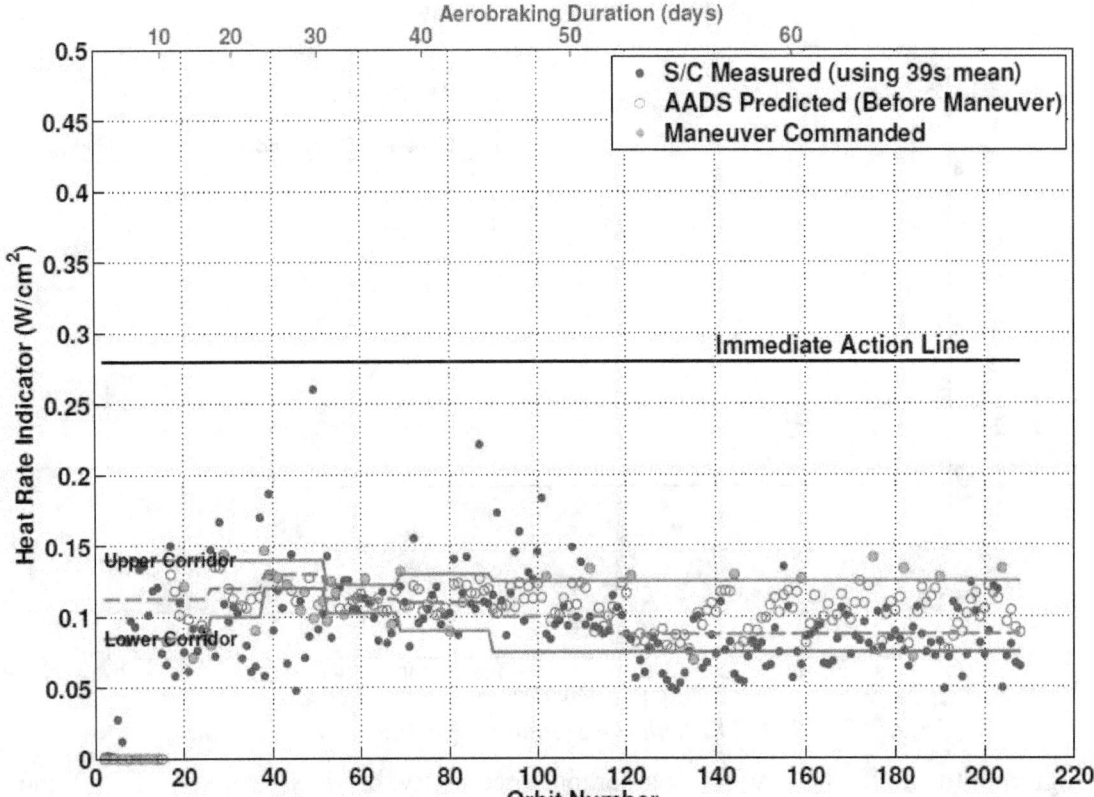

Figure 7.6-9. ORT2a with the Density 3-Sigma Correction

Finally, for comparison to the AADS ORT2 and ORT2a, which modified the corridor throughout the mission, a third AADS ORT simulation was considered that kept the upper and lower corridor constant for the entire mission (ORT2b). The corridor was set to the values determined from the operational mission run-out (0.085 and 0.14 W/cm^2). With the 3-sigma density maneuver logic enabled, ORT2b also had no immediate action violations. A full ORT mission comparison is provided in the next section.

For comparison, a summary of the ORT1 and ORT2 process is shown in Figure 7.6-10.

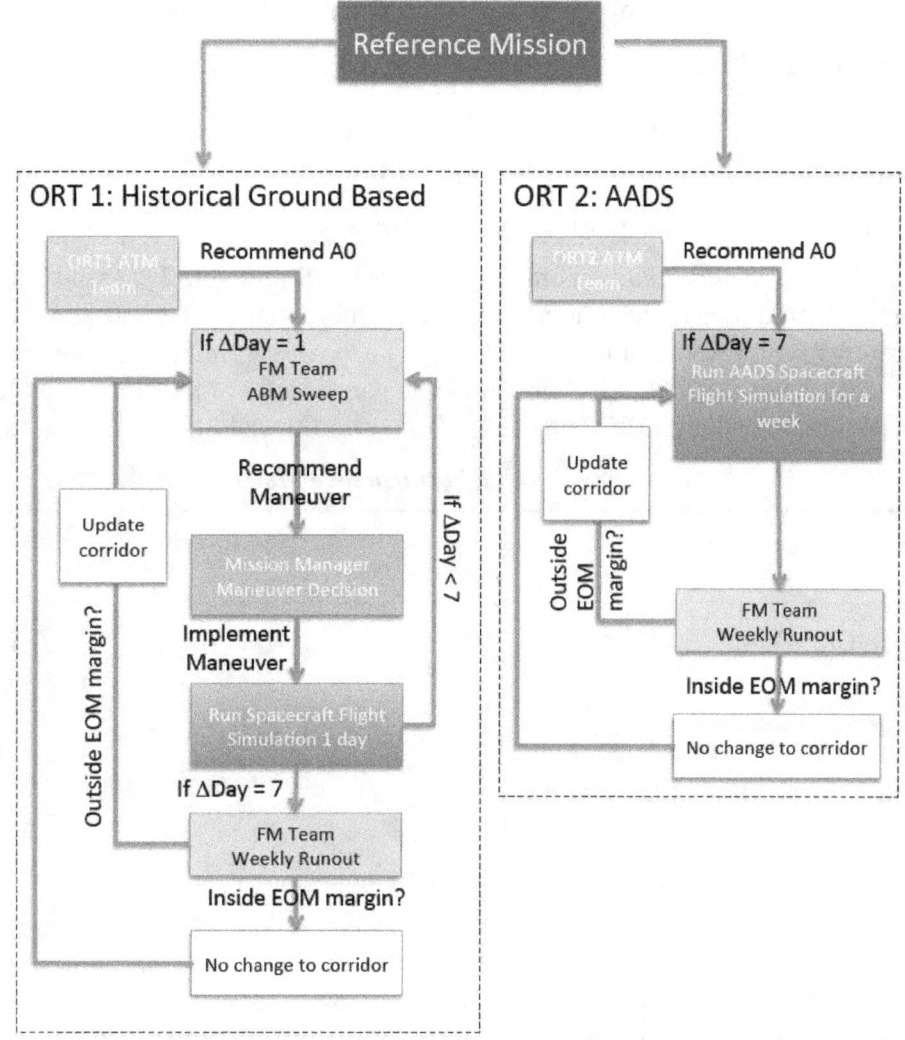

Figure 7.6-10. Data Flowchart for the Two Types of ORTs; EOM

7.6.1.5 ORT Performance Comparison

As was expected, the AADS ORT (ORT2) required much less ground support (both in terms of time and effort) and ended with a LTST closer to the reference mission than the ground-based ORT (ORT1). However, the required amount of ground support is a qualitative metric and is difficult to measure in the AA Phase 2 simulated and condensed ORT environment. Therefore, quantitative metrics are identified to compare the overall aerobraking mission performance between the ORTs.

Primary metrics used for comparison of the operational approaches included:

1. Total maneuver ΔV required
2. Total number of maneuvers
3. Flight corridors
4. Ability to match reference mission end conditions (e.g., LTST)

Table 7.6.2 compares key aerobraking parameters in each ORT. In ORT2a, which includes the density 3-sigma uncertainties in the maneuver logic, 23 more maneuvers and 75 percent more ΔV (15.3 versus 8.75 m/s) were required than for ORT2, however, no immediate action violations occurred. The corridor selections in ORT2a are similar to those used in the original ORT2, but the overall mission profile is different because of the additional biasing maneuvers.

Table 7.6-2. ORT Performance Metrics

Simulation	Number of Maneuvers	Maneuver ΔV (m/s)	Duration (days)	Final LTST
ORT1 – Ground	22	4.8	70.4	10:39
ORT2 – AADS	33	8.75	69.9	10:43
ORT2a – AADS (3 sig atm)	56	15.3	67.3	10:49
ORT2b – AADS (const corridor)	26	11.1	65.5	10.51

Figure 7.6-11 shows the variations in the flight corridors throughout the aerobraking mission for each ORT. ORT1, ORT2, and ORT2a modified corridors weekly based on the weekly run-out analysis and were adjusted to keep the mission on track with the reference glide slope to end at the desired LTST. ORT2b maintained a constant corridor throughout the mission

ORT2b required fewer maneuvers and used less ΔV than either of the other two AADS mission simulations, while still only requiring weekly ephemeris updates. It should be noted here that even with a fixed corridor, AADS still meets the final LTST requirements (i.e., target +/- 15 minutes). What is lost in the case of flying a fixed corridor for the entire mission is the ability to catch up in the event that the spacecraft falls behind, or provides more margin if the spacecraft has become ahead.

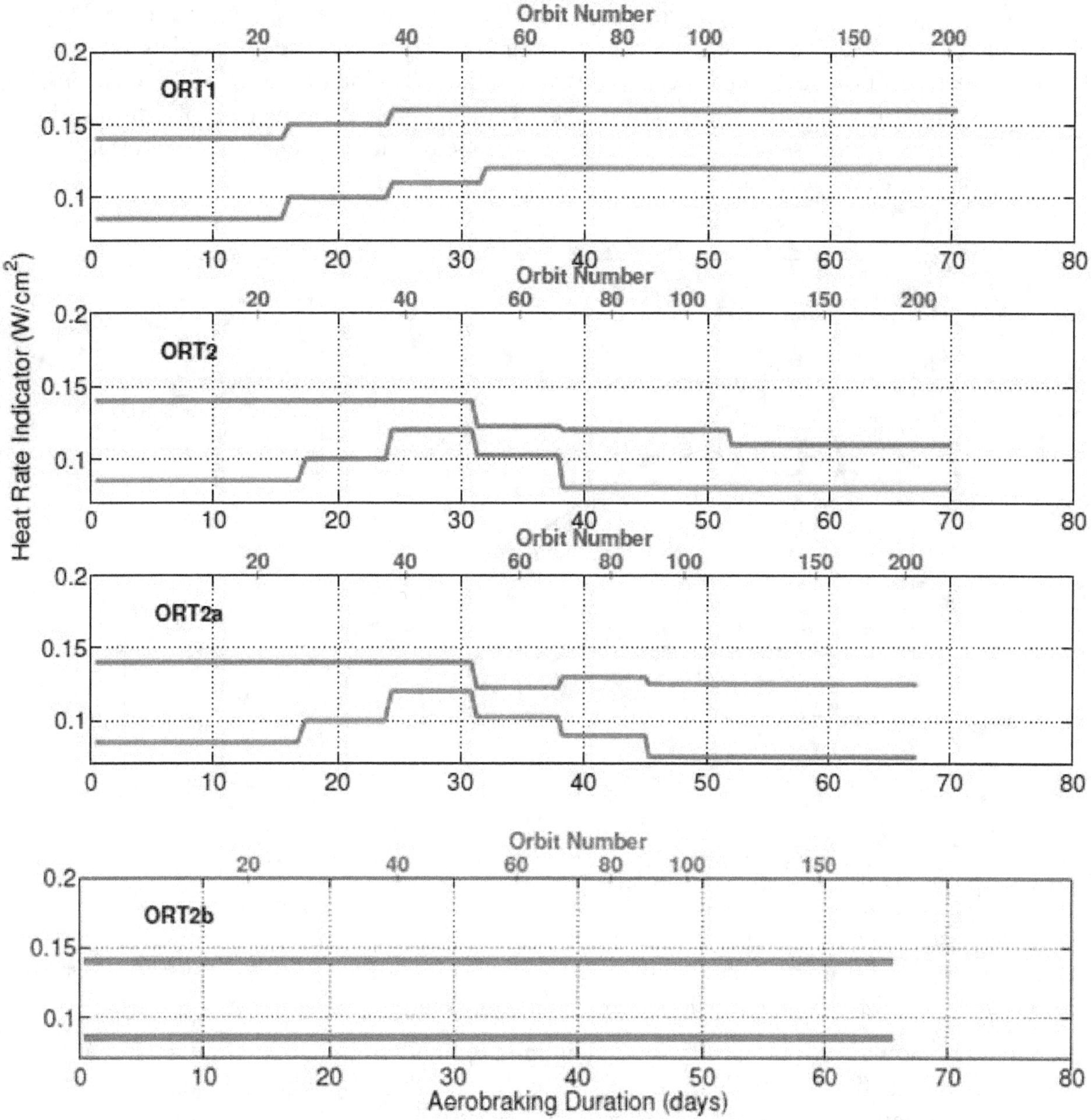

Figure 7.6-11. ORT Corridors

Thus, AADS provides both better control of the final LTST and more flexibility primarily because a maneuver can be performed every orbit.

Plots of the glide slope for the reference mission, ORT1 and ORT2 are shown in Figure 7.6-12. ORT2 was able to follow the reference glide slope more closely than the ground-based ORT1. Both ORTs start off a little behind the glide slope (on the right side of the reference mission), but AADS has more opportunities to take corrective action to get back on track with significantly less analysis and enabled several other corridor and maneuver logic scenarios to be tested.

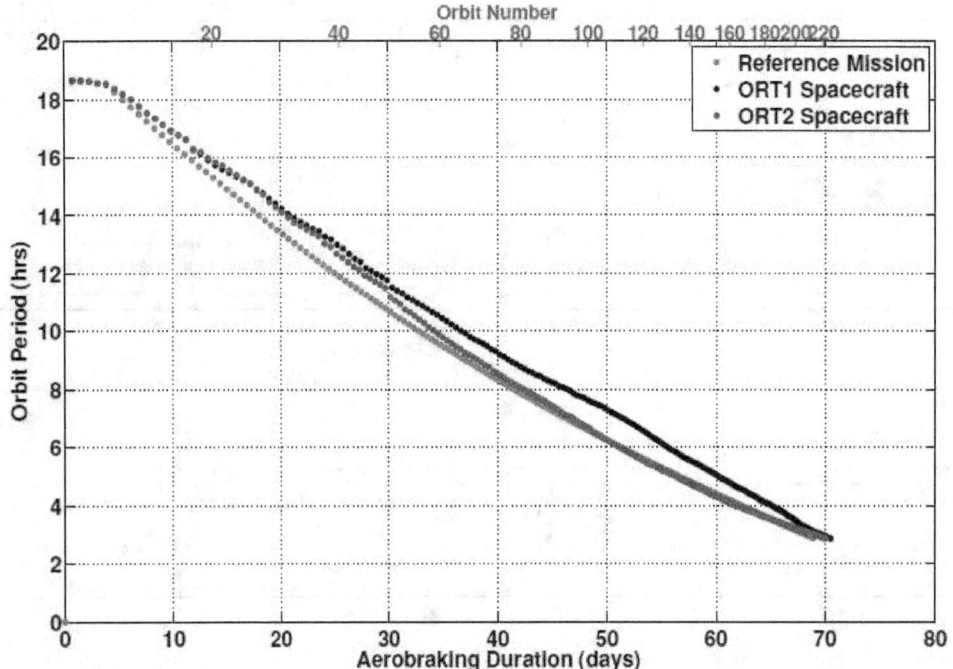

Figure 7.6-12. ORT Glide Slope Comparison to Reference

Figure 7.6-13 provides a comparison of the local time of the equatorial node crossing. As mentioned, the target was to end within 15 minutes of the reference mission. All ORT simulations were able to meet the requirement. Finally a plot of the local time versus orbit period for each ORT compared to the reference is provided in Figure 7.6-14.

The analyses shown here represent only the beginning of understanding the full set of AADS capabilities and its fundamentally different approach to aerobraking when the bulk of the maneuver decision-making is moved onboard the spacecraft.

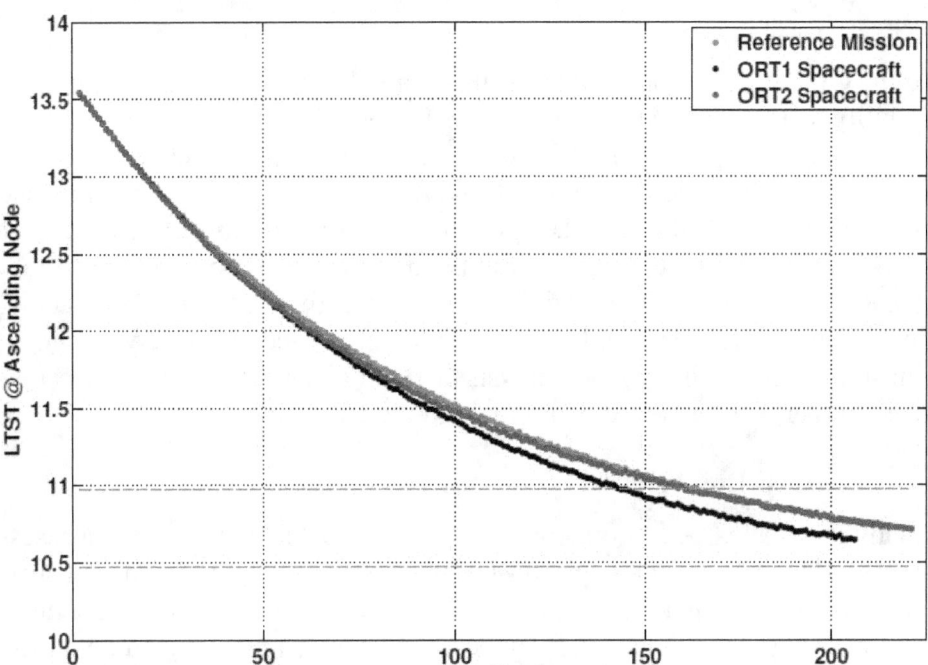

Figure 7.6-13. *LTST of the Ascending Node Comparison to Reference*

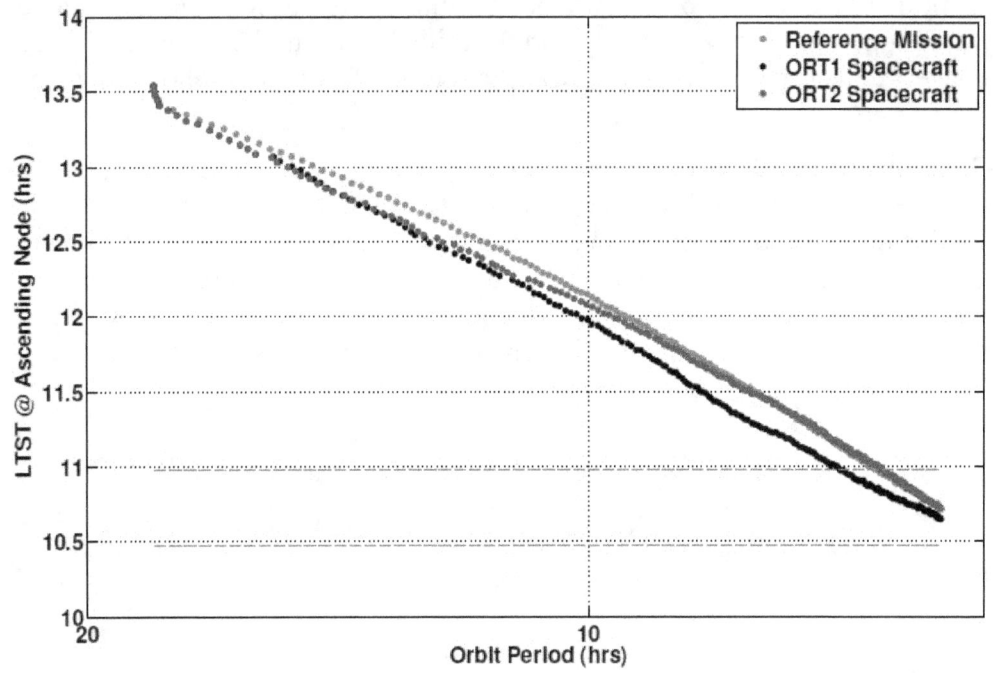

Figure 7.6-14. *ORT LTST versus Orbit Period Comparison to Reference*

Dust Storm Stress Case

In addition to the ODY type mission ORTs, a stress case "mini" ORT was performed to evaluate the AADS's ability to accommodate a dust storm. The MGS mission encountered a global dust storm early in its aerobraking mission. Because all the flight reconstructed density profiles were available, both ground-based and AADS "mini" ORTs were run using MGS atmosphere profiles corresponding to MGS orbits 50 to 70. The ground-based operations did experience some increase in workload due to higher variability in the day-to-day aerobraking performance. Compared to ORT1, there was increased difficulty in correctly predicting and reacting to large differences to maintain glide slope while trying to preserve spacecraft safety margin. However, there was almost no effect to the ground analysis or flight observed for the AADS operations due to its capabilities to perform maneuvers to remain in the corridor.

7.6.1.6 Lessons Learned/Bug Fixes

There were many lessons learned throughout the week the ORT took place. An assessment of the overall workload revealed that the ground-based operations required constant spacecraft monitoring and analysis. During the ORT, this task required 3 to 5 people. In contrast, the AADS simulation required much less staff support (one person), yet yielded better final mission conditions, namely the LTST of the ascending node crossing, at the end of aerobraking, while not requiring the thermal margin for the spacecraft to be reduced to compensate for falling behind. The primary reason the AADS mission minimizes the chances of falling behind is that the AADS can support a maneuver on every orbit, while the ground-based analysis could only perform a maneuver that would nominally keep the spacecraft safe for 48 hours (maximum being once daily).

Table 7.6-3 lists the issues identified in one or both of the ORTs during the entire process of planning and executing these ORT analyses as a way to evaluate the AADS performance and robustness. This table also includes the impact and resolution for each issue. It should be noted that all of the issues listed here were found to exist in the ORT analysis setup, truth or spacecraft simulation, or in external processing scripts and not within the AADS itself. This greatly increases the confidence in the design, implementation and interfaces of the AADS system, and indicates it is viable for future development.

Table 7.6-3. Issues Related to ORT Simulation and their Impacts and Resolutions

Issue	Effect	Impact	Rationale	Resolution
ODY atmosphere profiles were not being incremented due to error in table indexing	MAJOR	BOTH	Resulted in same atmosphere profile being used for each orbit, eliminating desired orbit-to-orbit variability	Identified during first week of ORT; was corrected and the ORT analyses restarted; walk-in trajectory was NOT corrected
20 orbit offset between ORT simulation pass number and ODY atmosphere profile (e.g., ORT pass #87 used ODY orbit 107 atmosphere profile)	MINOR	BOTH	Orbit-to-orbit variability is still preserved	No resolution was necessary for ORT; correction was implemented for ODY "re-flight" analyses
Atmosphere profiles were being changed at periapsis	MAJOR	AADS	Resulted in a different ODY profile being used for pre and post periapsis during any single drag pass; introduced atmosphere discontinuity at periapsis	Identified at end of ORT day 1; was corrected and AADS ORT mission and analyses from day 1 were re-run on day 2
Tolerance too small to allow for proper convergence of optimization problem in mission run-out simulations (mid-to-small period orbits)	MINOR	BOTH	Impact only to mission run-out analyses; issue was identified and corrected prior to the point in the ORT where any possibly corrupt data would have been used for weekly reset decisions	Tolerance was increased to allow for better convergence; this was done mid-ORT; re-runs of prior mission run-out analyses were not deemed necessary
Using time since periapsis instead of time to periapsis at start of drag pass to query ODY atmosphere profile	MINOR	AADS	Impact only to very beginning of drag pass (~200 km altitude) where the atmosphere profile table begins; result was incorrect interpolation of very small density; did not likely impact acceleration data into AADS, which begins/ends at a lower altitude (~180 km)	Required code change to AADS simulation (truth) was incorporated mid-ORT; re-run of prior analyses were not deemed necessary due to insignificant effect on truth trajectory

NESC Request No.: TI-09-00605

Issue	Effect	Impact	Rationale	Resolution
MarsGRAM inputs in target.inp inconsistent with those in msn_runout.inp, which resulted in different atmospheres being used in the optimization problem of the mission run-out analyses when the periapsis altitude became small (near EOM for low period orbits)	MINOR	BOTH	Impact only to mission run-out analyses and reference mission; issue was identified and corrected prior to the point in the ORT where any possibly corrupt data would have been used for weekly reset decisions	Inputs corrected mid-ORT; re-runs of prior mission run-out analyses were not deemed necessary; impact on reference mission expected to be small so no update made during ORT; reference updated for relevant post-ORT and reflight analyses
Tracked instantaneous heat rate instead of a running mean to determine maximum spacecraft thermal environment during a drag pass	NEUTRAL	BOTH	Due to the high "noise" level in the atmosphere data, the maximum heat rate based on the instantaneous density likely does not reflect the true spacecraft thermal environment; only impact was on the reporting of truth environments during weekly resets; since thermal limit violations were not significant, the impact was minimal	Simulations and post-processing scripts were updated to track the running mean heat rate to determine "true" spacecraft environment; implemented and used for all relevant post-ORT and reflight analyses
Atmosphere profiles misaligned with periapsis numbers; it was assumed that the atmosphere profile numbers were synced to the periapsis/orbit number when in fact it appears they are synced to the orbit number at the start of the pass	MINOR	BOTH	Impact on ORT analyses is insignificant; desired ODY orbit-to-orbit variability was preserved	Issue corrected for MRO reflight analyses; ODY reflight analyses were not correct since any impact on those results are assumed to be minor

Issue	Effect	Impact	Rationale	Resolution
startup atmosphere archive for AADS (AE) indexing error for density "predictions" of next orbit	NEUTRAL	AADS	Results in uncharacteristically high density sigma estimates when using the startup archive data (first 7 orbits of AADS operation; the effect is largest on orbit 1, minimal on orbit 7, and non-existent orbits ≥8); only impact is an increased possibility of commanding a bias maneuver when unnecessary (occurred only once in post-ORT analyses)	Startup atmosphere archive script updated for reflight analyses

Additional lessons learned included the need to establish a precise orbit number to stop the SFS. For ORT1, the code was set up to stop at the end of every 24-hour period, then wait to see if a maneuver would occur or not and if so, define the specified orbit. However, in hindsight, it would have been better to have stopped the simulation on an orbit number, a rather straightforward modification. This would have allowed for fewer maneuver options considered in the ABM sweep and likely implemented.

Also, after performing the ORT, as was found during ODY operations, having more options in the ΔV maneuver menu would have allowed more flexibility in selecting the most efficient maneuvers to remain in the corridor.

7.6.2 Historic Aerobraking Missions Replicated

In addition to the ORT analysis, a POST2 simulation was developed to evaluate the ability of AADS to replicate past missions that used ground-based operations. The objective was to compare the operational decisions that would have been made by AADS to the decisions made by the actual flight team. Two simulations were established, one for ODY and another for MRO. These aerobraking missions had different mission designs. ODY was an aggressive mission in that to meet the science requirements it had to fly with only a 100 percent heat rate margin (during the mission, this was reduced to 80 percent because it was flying too far behind the desired glideslope). MRO was a conservative mission with a heat rate margin of 250 percent. Another difference was that ODY encountered an atmosphere density that was ~20 percent of the nominal predicted whereas MRO encountered an atmosphere density that was ~500 percent of the nominal predicted. Thus, ODY had to fly lower periapsis than predicted and MRO flew higher periapsis than predicted.

Each mission utilized the respective aerobraking initial state, spacecraft mass prosperities and aerodynamics, and observed atmospheres from the actual mission. For simplification, the actual flight corridors were assumed to be constant for the analyses presented here, although additional cases were run to illustrate the AADS optimization capability demonstrated during the ORT analysis. The initial state was taken prior to the start of the walk-in phase of each respective mission. The walk-in phase was then manually duplicated (ensuring both the periapsis altitudes and heat rates were consistent between the actual mission and the AADS simulation) prior to the start of the main aerobraking phase, where the AADS is utilized. It is also noted that, since AADS is not intended for use in the last phases of aerobraking (endgame and walk-out where both COLA and spacecraft lifetime operations take precedence), both simulations are terminated for this analysis at an apoapsis altitude of ~2800 km rather than the required 400 km final apoapsis altitude. The comparisons of metrics (e.g., number of maneuvers, ΔV, mission duration) only include the historical data to the same point in the mission. The following sections describe the results of the AADS to replicate the ODY and MRO missions.

7.6.2.1 AADS ODY Aerobraking Mission Replication

The ODY aerobraking mission, following Mars orbit insertion on October 24, 1999, was the most aggressive aerobraking mission to date, meaning it had the lowest amount of margin to its thermal limits. ODY utilized a non-constant heat rate corridor that was initially set with an upper limit of 0.32 W/cm^2 and a lower corridor of 0.14 W/cm^2. The immediate action line was 0.54 W/cm^2. This corridor reduced the 18-hour initial orbit period to 2 hours over 77 days. Figure 7.6-15 shows the maximum heat rates observed each orbit of the ODY mission. To simplify the AADS evaluation, the ODY initial corridor values were held constant in the simulation. Figure 7.6-16 shows the results AADS following a similar corridor. As mentioned, the AADS is only intended for use during the main phase of aerobraking and is terminated early in the endgame near orbit 210. Table 7.6-4 provides a comparison of the AADS flight metrics to the actual mission values to that point in the mission. The AADS simulation required 10 more maneuvers, but used slightly less ΔV and ended only a fraction of a day earlier while requiring updates only weekly compared to the ground operations that occurred daily. The simulation verified that AADS was able to meet aerobraking requirements through the main phase.

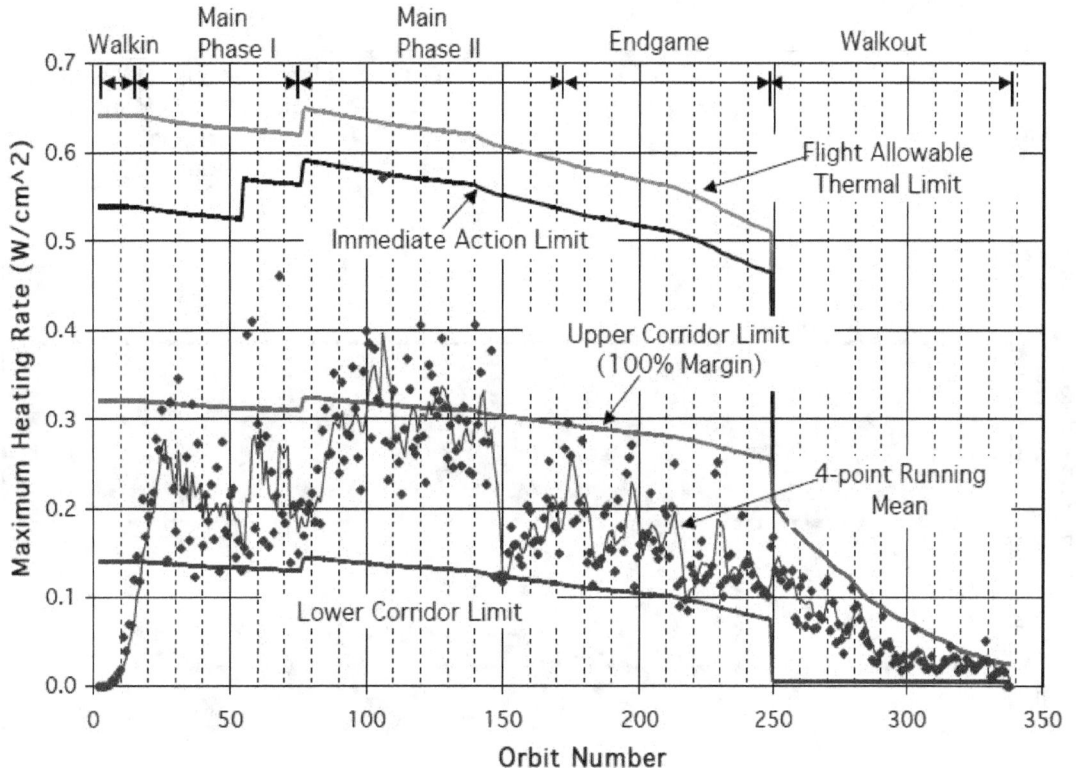

Figure 7.6-15. ODY Aerobraking Maximum Heat Rate [ref. 4]

Table 7.6-4. Comparison of ODY Flight and AADS Replicated Missions

Simulation	Number of Maneuvers	Maneuver ΔV (m/s)	Duration (days)	Change in final LTST of node (min)
ODY Flight	25	18.84	68.8	--
AADS fixed corridor	35	17.6	68.0	8.0
AADS small corridor	173	32.5	64.5	-0.33

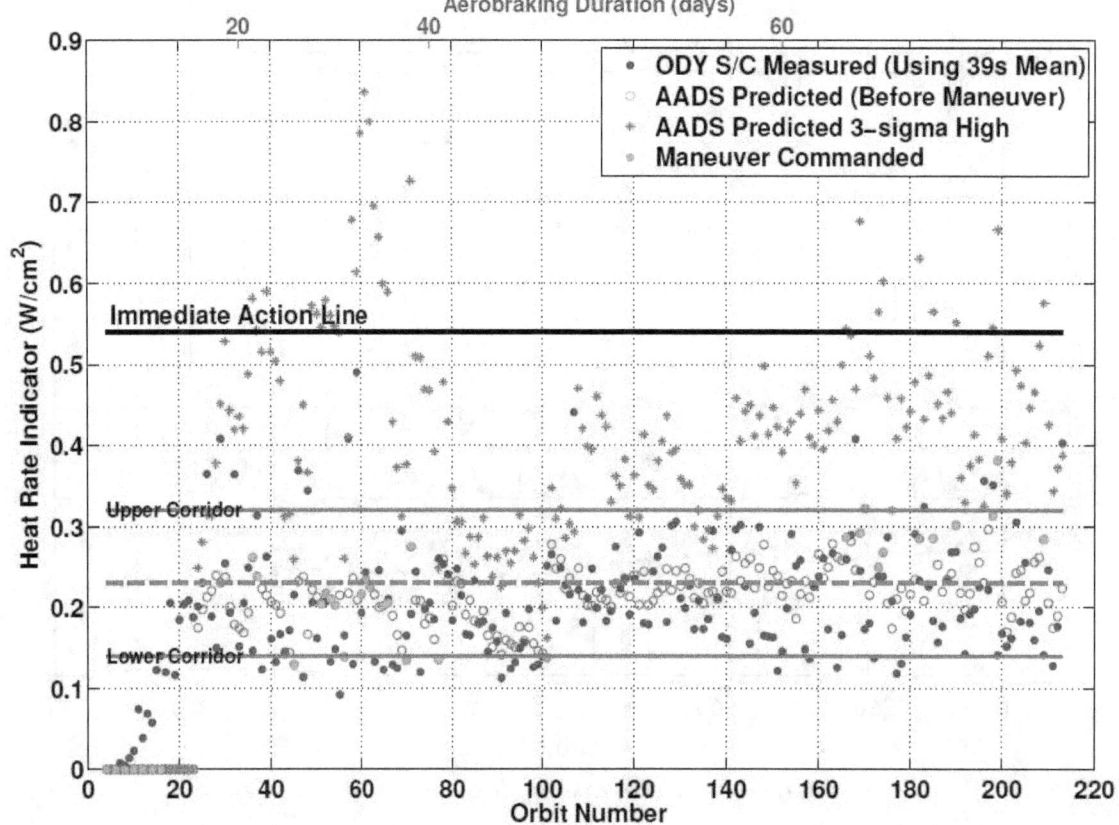

Figure 7.6-16. AADS ODY Aerobraking Mission Run-out Maximum Heat Rate

The AADS mission matched closely to the aerobraking duration and ΔV, but it missed the desired target LTST of the descending node by about 8 minutes (requirement was 15 minutes) and is shown in Figure 7.6-17. During the ODY aerobraking phase, ODY fell behind the desired glide slope to the point that it might not have met the required LTST requirement. ODY flight managers reduced the margin to 80 percent to allow ODY to fly higher heat rates to catch up. With AADS, the original margin could be maintained by raising the lower corridor, which would result in ODY flying higher in the corridor at the expense of more maneuvers. To demonstrate this, a second simulation was run with the corridor collapsed to always command the middle of the original corridor at 0.23 W/cm². Targeting the middle of the original corridor met the science requirements, but at the cost of 150 more maneuvers and twice the ΔV. Thus, if the spacecraft was using AADS and got behind (e.g., if there was a malfunction and the spacecraft had to increase its periapsis altitude while the problem was being resolved) AADS could be configured to fly higher in the corridor and catch up without having to reduce the thermal margin. For reference, Figures 7.6-18 and 7.6-19 show how the two AADS missions compared to the actual ODY flight with regard to periapsis latitude, altitude, orbit period, and LTST.

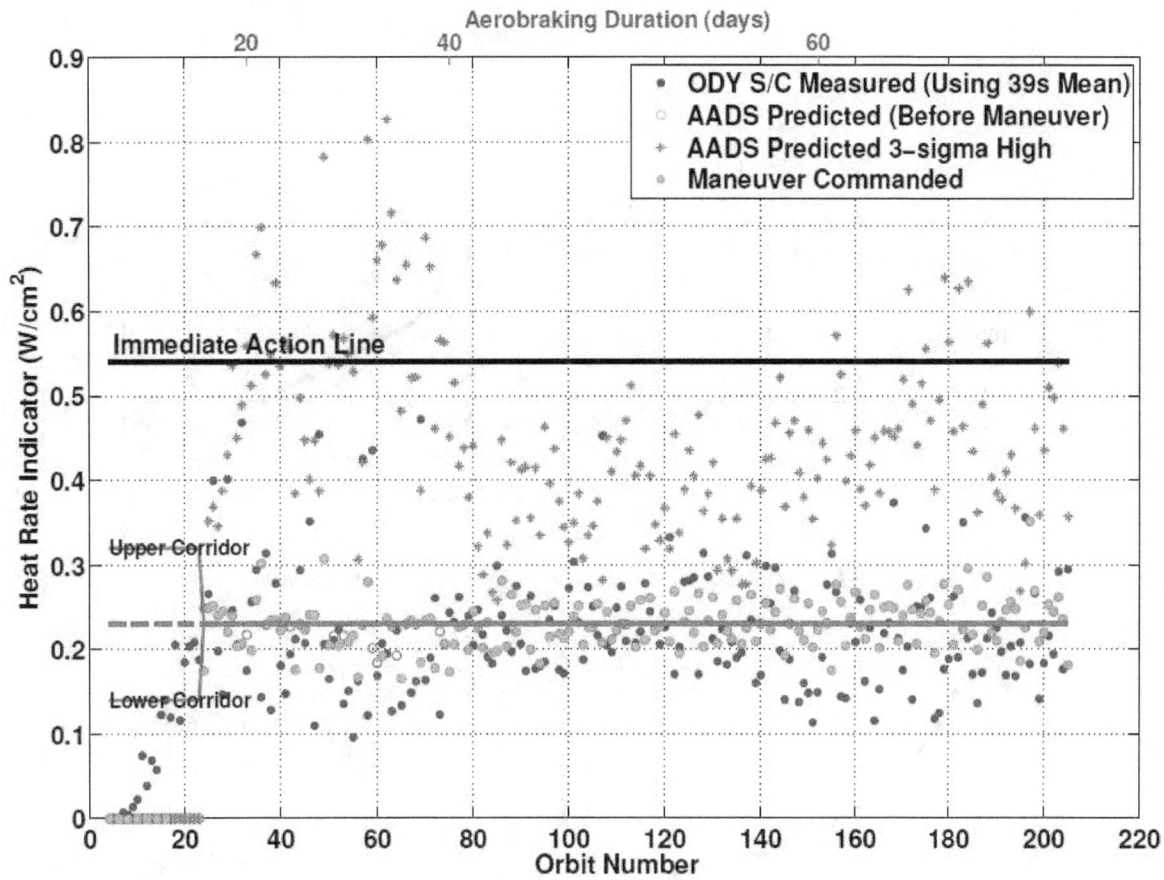

Figure 7.6-17. AADS ODY Narrow Corridor Simulation Peak Heat Rates

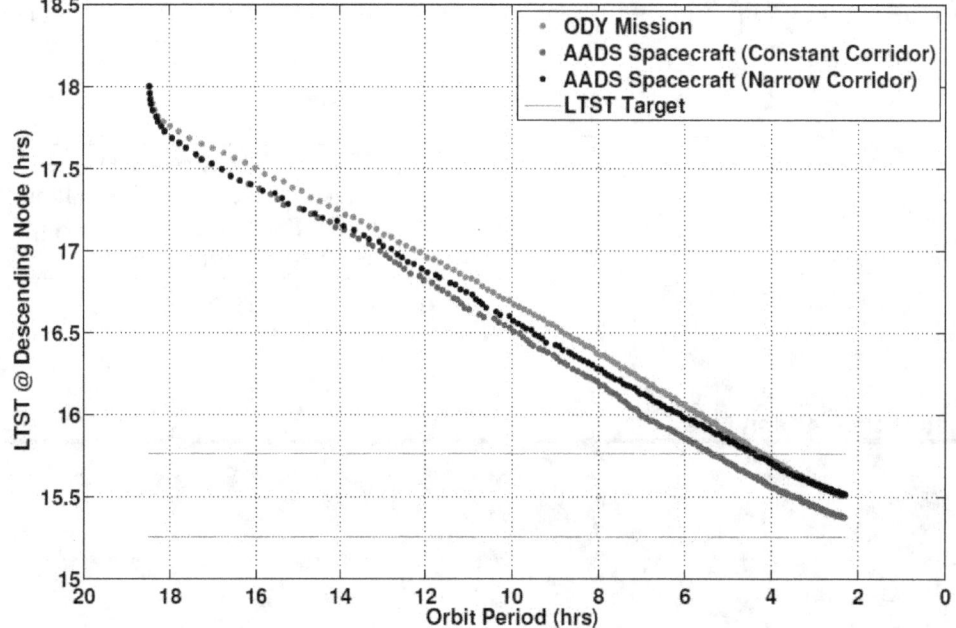

Figure 7.6-18. *ODY Mission versus AADS Glide Slope Comparison*

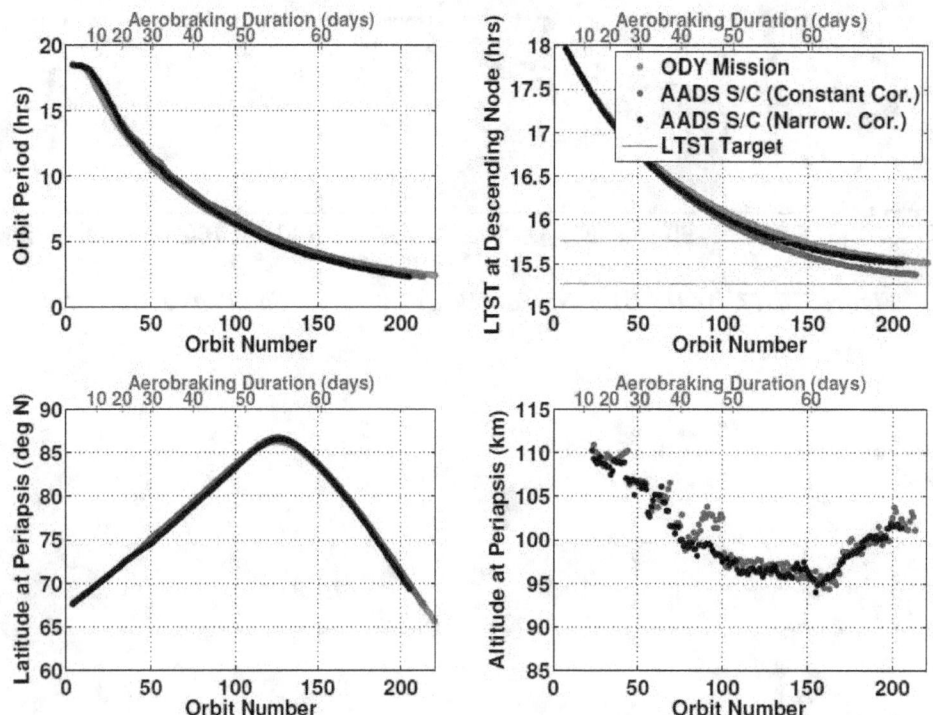

Figure 7.6-19. *ODY Mission Compared to AADS Two Replication Simulations*

7.6.2.2 AADS MRO Aerobraking Replication

The much less aggressive MRO aerobraking mission began on the 24th orbit past Mars orbit insertion on March 10, 2006. MRO had much lower corridor constraints with an average upper limit of 0.18 W/cm^2 and approximate lower corridor of 0.11 W/cm^2. The immediate action line was near 0.4 W/cm^2. Figure 7.6-20 shows the maximum heat rates observed on each orbit of the MRO mission. Again, to simplify the AADS evaluation, the initial corridor values were held constant in the simulation. Figure 7.6-21 shows the results AADS following a similar corridor. Table 7.6-5 provides a comparison of the AADS flight metrics to the actual mission values. The AADS simulation required 18 more maneuvers requiring 8.4 m/s ΔV and ended 10 days later than the original mission. As was shown for the ODY mission, the AADS was able to meet the MRO aerobraking mission requirements using spacecraft updates weekly compared to the daily ground operations that were required for the actual mission. See Figures 7.6-22 and 7.6-23.

Table 7.6-5. Comparison of MRO Flight and AADS Replicated Mission

Simulation	Number of Maneuvers	Maneuver ΔV (m/s)	Duration (days)	Change in final LTST of node (min)
MRO Flight	24	16.0	157.2	--
AADS	42	24.4	161.0	11.3

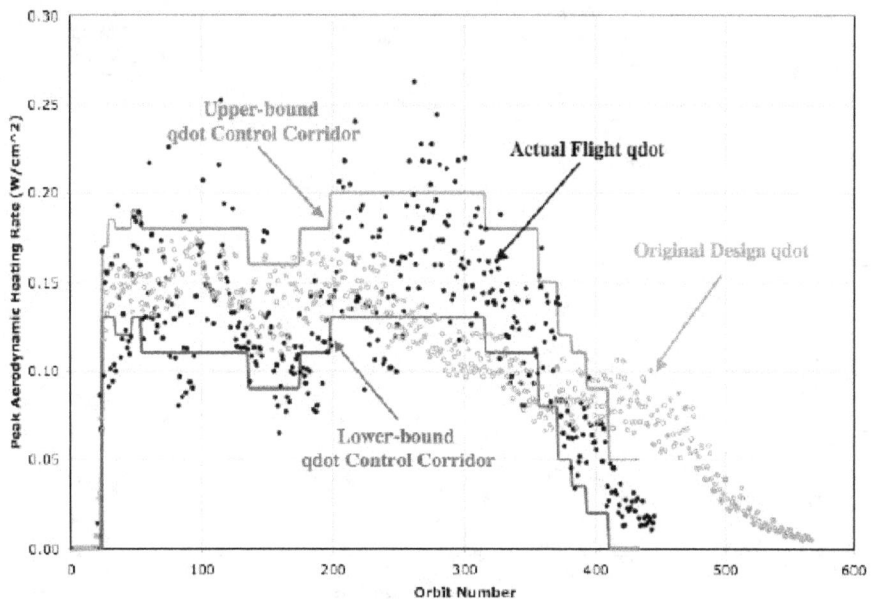

Figure 7.6-20. MRO Aerobraking Peak Heat Rates [ref. 5]

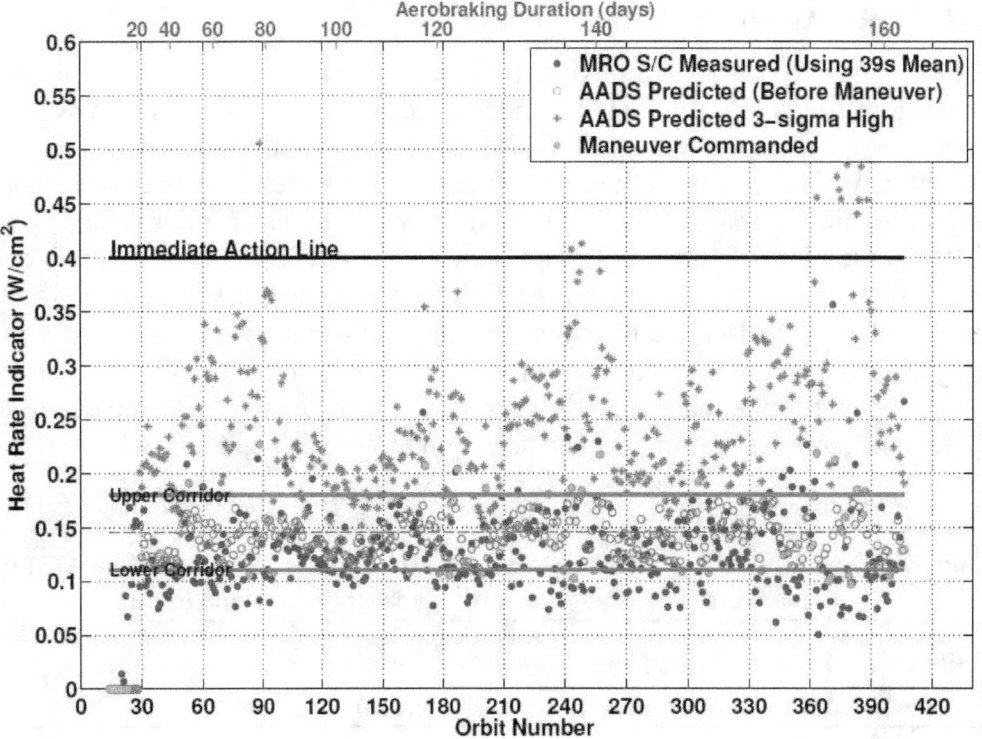

Figure 7.6-21. *AADS MRO Aerobraking Mission Run-out Maximum Heat Rate*

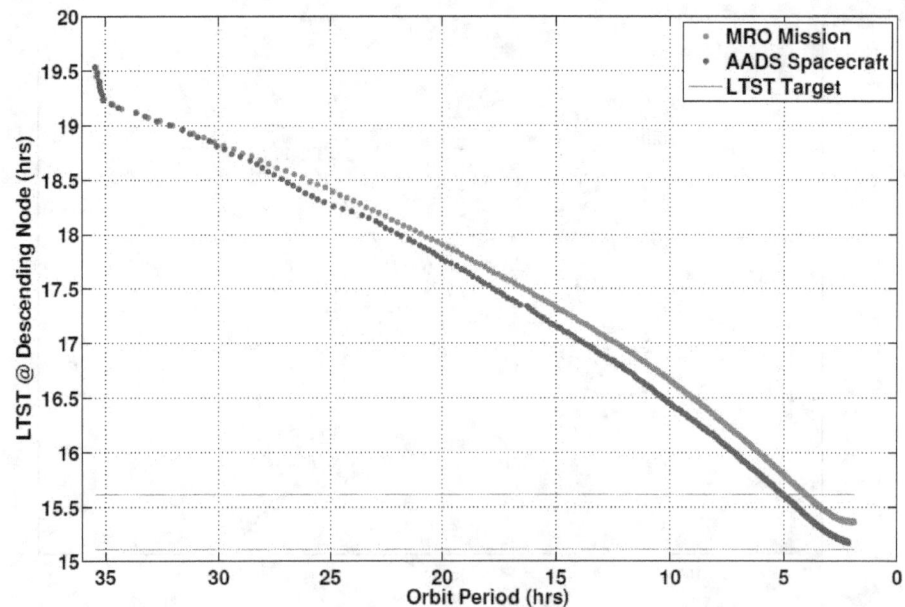

Figure 7.6-22. *MRO versus AADS Glide Slope Comparison*

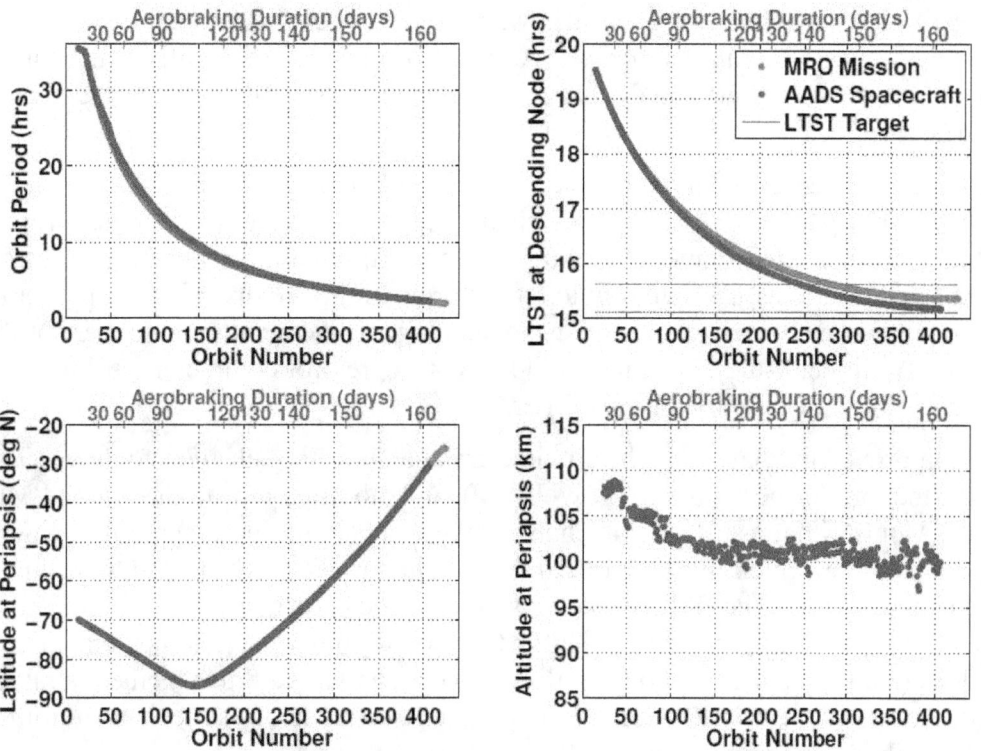

Figure 7.6-23. MRO versus AADS Aerobraking Mission Parameter Comparison

The purpose of evaluating the various AADS operational scenarios was to understand the features and capabilities of the software. The cases examined for the AA Phase 2 ORTs are the beginning to understanding the flexibility and capability of AADS. By implementing the biasing, the number of maneuvers increased (ORT2a) and relaxing the corridor by keeping it constant throughout the mission (ORT2b) the mission used fewer maneuvers with a shorter aerobraking phase. Different options could be considered, such as only using the biasing during the periods with the most atmosphere variability. Yet, none of the stress cases resulted in a modification of the AADS code only in how it was implemented in the mission. The ORT demonstrated the power of having the onboard maneuver capability. AADS is the logical extension of the current periapsis timing estimator implemented on ODY and demonstrated on MRO and is a means to enable a new era of planetary exploration in the solar system.

7.7 Open Phase 2 Investigation Topics

The AA Development Team completed an extraordinary amount of work during Phase 2. Simulations that were built in Phase 1 were developed and improved during Phase 2. Many trade studies were performed; nominal, off-nominal, and stressing aerobraking scenarios were analyzed, and Monte Carlo performance was assessed. In addition, the team developed and executed an ORT of an aerobraking scenario with a stressing environment and recreated ODY-like and MRO-like aerobraking missions. AADS performed well under all circumstances. There are, however, areas of continued study that could be accomplished during a Phase 3 task. These items are described below.

Phase 2 focused on the development of AA at Mars. Although some work was performed in the area of Venus atmospheric uncertainty analysis [ref. 7], few simulations were performed to study the effect of a stressing environment at Venus. Given additional time and resources, trade studies and Monte Carlo assessments could be performed for a Venus-orbiting spacecraft. In addition, an ORT-like scenario could be developed if flight-like Venus atmosphere profiles could be produced. Additional studies at Titan could be considered for completeness, but it is a less probable destination for an aerobraking orbiter.

AADS was improved in Phase 2 by the architecture and algorithm changes to the orbit integration function (formerly the EE, now the SP), and modifications to the AE and ME algorithms. Along with these algorithm changes, the software has been streamlined with the EML implementation, which further reduces the computational overhead and the memory requirements for AADS. The EML version also moves the code closer to a version suitable for implementation on a flight processor, although there is still much work to be done in this area. An error detection strategy, exception handling strategy, and a fault management strategy need to be defined. Some will be mission-dependent, but the software needs a mechanism for reporting errors and possibly responding to them.

Another issue identified in Phase 2 was the characterization of the uncertainty of the control variable. When the heat rate is used for the control variable, the AADS computes the uncertainty (variance) associated with the atmosphere model fit to help inform the maneuver calculation algorithm. This variance can be used to decide whether or not to perform a second category of maneuver, which might help further ensure spacecraft safety (instead of simply attempting to follow the designed glide slope of the reference mission). While it is largely believed that the natural variation in the atmosphere is the dominant source of uncertainty in the corridor control variable, there are other uncertainty sources that should be characterized for a more thorough understanding of the robustness of AA. Of particular importance are the uncertainties due to the orbit propagation. Because this process is only reset via ground interaction, this largely determines the duration between ground updates. Initial state uncertainty (in the ground-based navigation solution) and uncertainties in the accelerometer data are generally small, but grow large through the long orbit propagation arcs. This can lead to uncertainty in periapsis altitude and timing predictions which, in turn, affect the atmosphere estimation and corridor control variable predictions.

A final issue identified in Phase 2 is related to the ground update. In current testing, the simulations use a current state provided to the spacecraft to reset the onboard ephemeris knowledge. This instantaneous state update is unrealistic as there would likely be substantial time between the ground computation of the orbit and the ability to load a new ephemeris state to the spacecraft. This latency may mean that several drag passes have transpired since the end of the ground tracking data arc, which degrades the orbit knowledge considerably, as the ground does not have an accurate atmosphere model for propagation. In this case, the spacecraft has better information about the recent past than the ground. In order for the ground to provide the spacecraft the most accurate state, this state may have to be several orbits old, which then requires the spacecraft to propagate this delayed state measurement up to the current time. This propagation may span numerous drag passes that require some method of storage or characterization to preserve the accuracy in the ground update. This problem is currently under investigation by the team, but will be challenging to solve without introducing substantial code changes or new concepts of operations.

Even with these identified studies to improve performance of AADS, Phase 2 provided key enhancements to the AADS and answered numerous important questions. In addition to the tasks assigned for Phase 2, an attempt was initiated to recreate the AADS ORT within the AAHFS. This was beyond the scope of the initial task, but much progress was made in performing this analysis.

A significant objective of Phase 2 was to provide two working versions of AADS: one using a SP developed by Applied Physics Laboratory (APL), the second using a version of the JPL-developed AeroNav. All of the analysis shown in this final report was using the SP version of AADS. Work was performed in developing and delivering AeroNav to the AADS implementation team, but following a unit test of the performance of AeroNav, there was not enough time left in Phase 2 to complete the integration of AeroNav into AADS and run the analyses. Phase 2 was successful in completing a broad suite of studies with AADS-SP. The AeroNav version of the AADS was desired for ease of implementation within a robotic JPL-led orbiter mission that would already have been using AutoNav. It is still desired to maintain this option, and with further study, AeroNav could be implemented and tested with the same rigor that was given to the AADS-SP.

Any further trade studies that have been identified for the AADS (e.g., ΔV sizing, maneuver magnitude limitations, maneuver execution performance sensitivities, etc.) will be left for implementation for a specific mission. These small studies are too spacecraft-specific to provide much benefit for the general AA analysis performed.

8.0 Findings, Observations, and NESC Recommendations

8.1 Findings

The following findings were identified:

F-1. The AADS performance at Mars was similar to that of a ground-based team, illustrated by the ORT and by recreation of the ODY and MRO aerobraking missions.

F-2. The ORT demonstrated the daily work requirements utilized for current ground-based operations could be extended to every 7th day for AADS.

F-3. The ORT demonstrated an ephemeris update only once per week is sufficient to maintain successful aerobraking with the AADS, which reduces the requirement for DSN from continuous coverage to a few hours per week.

F-4. The ORT demonstrated that the aerobraking corridor could be reduced to require a maneuver every orbit with the SP maintaining the required accuracy.

F-5. The ORT demonstrated the ability to execute the required maneuvers to continue aerobraking with the required margins when subjected to a simulated dust storm equivalent to that encountered by MGS.

F-6. AA simulations demonstrated that the number of maneuvers could be traded for thermal margin.

F-7. AA simulations replicated the ODY and MRO missions with no reductions in thermal margins.

F-8. A unit test of the AeroNav orbit propagation showed similar results to AADS (i.e., state differences <10 seconds) when tested at two independent locations.

F-9. The AADS simulations demonstrated robustness to reconstructed atmosphere characteristics using MRO, MGS, and ODY flight data.

F-10. The AADS algorithms are feasible for an onboard implementation in terms of execution speed and memory requirements.

F-11. Benchmarking of the AADS for onboard implementation demonstrated that computational frame requirements could be easily met with significant margin.

F-12. The AADS internal architecture was revised to allow sequential processing of input sensor data, eliminating large data buffers.

F-13. The number and frequency of the gravity model evaluations was dramatically reduced, reducing the computational burden on the flight processor of AADS.

F-14. AADS SP errors were greatly improved over Phase 1.

F-15. The fidelity of the onboard orbital integration was significantly improved by separating the integration of the fast dynamics (thrust and drag) from the slow dynamics (gravity model terms) and using a numerically efficient method for each integration.

F-16. In IMU noise-free simulation, ground updates can be spaced greater than 28 days with no degradation in ephemeris performance.

F-17. Corridor performance is not significantly affected by typical maneuver execution errors.

8.2 Observations

The following observations were identified:

O-1. Maintaining two independent and synergistic simulations (POST2 and AAHFS) allowed for rapid, parallel trade studies performed in parallel, providing increased quantity of Phase 2 analyses.

O-2. The EE AeroNav is not as mature as the AADS.

O-3. AADS does not employ algorithms to address latency of ground-based ephemeris updates.

O-4. Venus atmospheric uncertainties are substantially different than those calculated for Mars and may have significant impact in Venus aerobraking.

8.3 NESC Recommendations

The following NESC recommendations were identified and directed toward the NESC (including the technical disciplines of Flight Mechanics, Aerosciences, Passive Thermal, GN&C, Software, Loads and Dynamics, and Human Factors) and future NASA Programs and Projects that may utilize aerobraking:

R-1. Future missions that require aerobraking should evaluate the application of AADS. *(F-1 through F-7, F-9 through F-11)*

R-2. The ORT simulation capability developed in the Phase 2 effort could be utilized by future aerobraking missions to offer mission designers, project managers and spacecraft operators a realistic flight mission rehearsal. *(F-1 through F-7)*

R-3. Further development of AA should address:

    a. Evaluate Venus AADS performance, including potential ORT analysis and off-nominal testing. *(F-1 through F-7, F-9, O-4)*

    b. Implement AeroNav in AADS, provide alternative algorithms for autonomous aerobraking and perform an ORT. *(F-8, O-2)*

c. Evaluate and improve the suitability of AADS for flight processor implementation, ensuring code robustness and proper error reporting. *(F-10 through F-17)*

d. Develop AADS algorithms to address the latency of ground-based ephemeris updates. *(O-3)*

9.0 Alternate Viewpoint

There were no alternate viewpoints identified during the course of this assessment by the NESC team or the NRB quorum.

10.0 Other Deliverables

No unique hardware, software, or data packages, outside those contained in this report, were disseminated to other parties outside this assessment.

11.0 Lessons Learned

No applicable lessons learned were identified for entry into the NASA Lessons Learned Information System.

12.0 NASA Standards and Specifications

No recommendations for NASA standards and specifications were identified as a result of this assessment.

13.0 Definition of Terms

Corrective Actions	Changes to design processes, work instructions, workmanship practices, training, inspections, tests, procedures, specifications, drawings, tools, equipment, facilities, resources, or material that result in preventing, minimizing, or limiting the potential for recurrence of a problem.
Finding	A conclusion based on facts established by the investigating authority.
Lessons Learned	Knowledge or understanding gained by experience. The experience may be positive, as in a successful test or mission, or negative, as in a mishap or failure. A lesson must be significant in that it has real or assumed impact on operations; valid in that it is factually and technically correct; and applicable in that it identifies a specific design, process, or decision that reduces or limits the potential for failures and mishaps, or reinforces a positive result.

Observation	A factor, event, or circumstance identified during the assessment that did not contribute to the problem, but if left uncorrected has the potential to cause a mishap, injury, or increase the severity should a mishap occur. Alternatively, an observation could be a positive acknowledgement of a Center/Program/Project/Organization's operational structure, tools, and/or support provided.
Problem	The subject of the independent technical assessment.
Proximate Cause	The event(s) that occurred, including any condition(s) that existed immediately before the undesired outcome, directly resulted in its occurrence and, if eliminated or modified, would have prevented the undesired outcome.
Recommendation	An action identified by the NESC to correct a root cause or deficiency identified during the investigation. The recommendations may be used by the responsible Center/Program/Project/Organization in the preparation of a corrective action plan.
Root Cause	One of multiple factors (events, conditions, or organizational factors) that contributed to or created the proximate cause and subsequent undesired outcome and, if eliminated or modified, would have prevented the undesired outcome. Typically, multiple root causes contribute to an undesired outcome.

14.0 Acronyms List

A0	MarsGRAM Density Multiplier
AA	Autonomous Aerobraking
AADS	Autonomous Aerobraking Development Software
AAHFS	Autonomous Aerobraking High Fidelity Simulation
ABM	Aerobraking Maneuver
AE	Atmosphere Estimator
AMA	Analytical Mechanics and Associates
APL	Applied Physics Laboratory
APP	Aerobraking Planning Period
CCD	Central Composite Design
COLA	Collision and Avoidance
CPU	Central Processor Unit
DOE	Design of Experiments
DOF	Degrees of Freedom
DSN	Deep Space Network
ΔV	Delta Velocity

EE	Ephemeris Estimator
EML	Embedded MATLAB
EOM	End of Mission
ET2000	Ephemeris Time from midnight January 1, 2000
GN&C	Guidance, Navigation, and Control
GRAM	Global Reference Atmosphere Model
ICRF	International Celestial Reference Frame
IMU	Inertial Measurement Unit
JPL	Jet Propulsion Laboratory
LaRC	Langley Research Center
LTST	Local True Solar Time
MarsGRAM	Mars Global Reference Atmosphere Model
MB	megabyte
ME	Maneuver Estimator
MHz	Mega Hertz
MRO	Mars Reconnaissance Orbiter
MTSO	Management Technical Support Office
N	number
NASA	National Aeronautics and Space Administration
NAV	Navigation
NCSU	North Carolina State University
NESC	NASA Engineering and Safety Center
NIA	National Institute of Aerospace
NRB	NESC Review Board
OD	Orbit Determination
ODY	Mars Odyssey
OP	Orbit Propagation
ORT	Operational Readiness Test
PDS	Planetary Data System
POST2	Program to Optimize Simulated Trajectories (v2)
RKN	Runge-Kutta-Nystrom
RSE	Response Surface Equation
SD	State Determination
SFS	Spacecraft Flight Simulation
SP	State Propagator
TDT	Technical Discipline Team
TRL	Technology Readiness Level
VML	Virtual Machine Language

15.0 References

1. You, T. et al. "Mars Reconnaissance Orbiter Navigation," AIAA 2004-4745. 2004.
2. Gladden, R. "Mars Reconnaissance Orbiter: Aerobraking Sequencing Operations and Lessons Learned," AIAA 2008-3353. 2008.
3. Murri, Daniel G. et al. "Development of Autonomous Aerobraking (Phase 1)," December 15, 2011. NESC-RP-09-00605/NASA-TM-2012-217328.
4. Smith, J. C., "2001 Mars Odyssey Aerobraking" AIAA 2002-4532, 2002.
5. Long, S. M., et al. "Mars Reconnaissance Orbiter Aerobraking Navigation Operation", AIAA 2008-3349. 2008.
6. Lyons, D. T., et al. "Mars Global Surveyor: Aerobraking Mission Overview," Journal of Spacecraft and Rockets, Vol. 36, No. 3, May–June 1999.
7. Tolson, R., Prince, J., Konopliv, A. "An Atmospheric Variability Model for Venus Aerobraking Missions." Control ID#: 1654639: Technical Paper has been accepted for presentation at AIAA Modeling and Simulation Technologies Conference, Boston, MA, August 19-23, 2013.

www.ingramcontent.com/pod-product-compliance
Lightning Source LLC
Chambersburg PA
CBHW081731170526
45167CB00009B/3775